Series in Real Analysis – Vol. 16

The Non-uniform Riemann Approach to Stochastic Integration

SERIES IN REAL ANALYSIS

ISSN: 1793-1134

Published

Vol. 16: *The Non-uniform Riemann Approach to Stochastic Integration*
Varayu Boonpogkrong, Tuan Seng Chew & Tin Lam Toh

Vol. 15: *Kurzweil–Stieltjes Integral: Theory and Applications*
Giselle Antunes Monteiro, Antonín Slavík & Milan Tvrdý

Vol. 14: *Nonabsolute Integration on Measure Spaces*
Wee Leng Ng

Vol. 13: *Theories of Integration: The Integrals of Riemann, Lebesgue, Henstock–Kurzweil, and McShane (Second Edition)*
Douglas S Kurtz & Charles W Swartz

Vol. 12: *Henstock–Kurzweil Integration on Euclidean Spaces*
Tuo Yeong Lee

Vol. 11: *Generalized Ordinary Differential Equations: Not Absolutely Continuous Solutions*
Jaroslav Kurzweil

Vol. 10: *Topics in Banach Space Integration*
Štefan Schwabik & Ye Guoju

Vol. 9: *Theories of Integration: The Integrals of Riemann, Lebesgue, Henstock–Kurzweil, and McShane*
Douglas S Kurtz & Charles W Swartz

Vol. 8: *Integration between the Lebesgue Integral and the Henstock–Kurzweil Integral: Its Relation to Local Convex Vector Spaces*
Jaroslav Kurzweil

Vol. 7: *Henstock–Kurzweil Integration: Its Relation to Topological Vector Spaces*
Jaroslav Kurzweil

Vol. 6: *Uniqueness and Nonuniqueness Criteria for Ordinary Differential Equations*
R P Agarwal & V Lakshmikantham

Vol. 5: *Generalized ODE*
Š Schwabik

Vol. 4: *Linear Functional Analysis*
W Orlicz

Vol. 3: *The Theory of the Denjoy Integral & Some Applications*
V G Celidze & A G Dzvarseisvili
translated by P S Bullen

Vol. 2: *Lanzhou Lectures on Henstock Integration*
Lee Peng Yee

Vol. 1: *Lectures on the Theory of Integration*
R Henstock

Series in Real Analysis – Vol. 16

The Non-uniform Riemann Approach to Stochastic Integration

Varayu Boonpogkrong
Prince of Songkla University, Thailand

Tuan Seng Chew
National University of Singapore, Singapore

Tin Lam Toh
Nanyang Technological University, Singapore

NEW JERSEY • LONDON • SINGAPORE • BEIJING • SHANGHAI • HONG KONG • TAIPEI • CHENNAI • TOKYO

Published by

World Scientific Publishing Co. Pte. Ltd.
5 Toh Tuck Link, Singapore 596224
USA office: 27 Warren Street, Suite 401-402, Hackensack, NJ 07601
UK office: 57 Shelton Street, Covent Garden, London WC2H 9HE

Library of Congress Control Number: 2024039536

British Library Cataloguing-in-Publication Data
A catalogue record for this book is available from the British Library.

Series in Real Analysis — Vol. 16
THE NON-UNIFORM RIEMANN APPROACH TO STOCHASTIC INTEGRATION

ISBN 978-981-98-0122-0 (hardcover)
ISBN 978-981-98-0123-7 (ebook for institutions)
ISBN 978-981-98-0124-4 (ebook for individuals)

For any available supplementary material, please visit
https://www.worldscientific.com/worldscibooks/10.1142/14052#t=suppl

Printed in Singapore

Preface

This book aims to present the stochastic integration theory using the (generalized) Riemann approach. It has always been said in classical textbooks on stochastic calculus that it is "impossible" to define stochastic integrals using the Riemann approach. This is because even the most fundamental stochastic process, the Brownian motion, does not have paths of bounded variation.

In this book, we define the stochastic integral using the *generalized* Riemann approach, which was independently discovered by Kurzweil and Henstock in 1950s in classical (non-stochastic) integration theory. The technically minor modification from the traditional Riemann approach has resulted in the definition of an integral which encompasses the Riemann and Lebesgue integral. The generalized Riemann approach, in which the uniform mesh of the Riemann approach is replaced by a mesh that varies from point to point, has resulted in its ability to integrate functions which are highly oscillating.

In this book, we use the generalized Riemann approach to define a stochastic integral. Our integral encompasses the classical stochastic integrable stochastic process. We consider the cases where the integrators are canonical Brownian motion, martingales, continuous semi-martingales, and even integration in higher dimensions. We highlight that the definition of the integral using the generalized Riemann approach gives an explicit definition and taps on the intuitive approach of the Riemann integral.

The Lebesgue integral, a generalization of the classical Riemann integral, utilises much measure theoretical tools in its definition. Such an approach may be a challenge for students who are non-specialist in mathematics. Along the same line of thought, Itô defines the stochastic integral with respect to a canonical Brownian motion, see [Itô (1944)], making use of simple processes and isometric equality (known as Itô isometry) and the completeness of L^2-norm in its definition. The generalized Riemann approach is able to offer an alternative definition which builds on the intuitive nature of the Riemann integral. We believe that our approach is pedagogically sound, especially for students who are non-specialist in mathematics, yet require the stochastic integral in other disciplines.

This book is an update and development based on [Toh (2001)] a doctoral dissertation submitted to the National University of Singapore. More than two decades from 2001, much development has occurred, including those found in [Boonpogkrong (2004); Boonpogkrong and Chew (2004); Boonpogkrong, Chew and Toh (Under Review); Chew, Huang and Wang (2004); Canton, Labendia and Toh (2022); Muldowney (2012); Tan and Toh (2011–12); Toh and Chew (2002, 2003–04, 2004–05, 2005, 2010, 2012); Yang and Toh (2014, 2016); Lim and Toh (2022)].

Chapter 1 provides an introduction to the whole book in several aspects: (1) a discussion of the generalized Riemann approach in the non-stochastic integration theory that has been used by various researches; (2) the properties of Brownian motion and martingale, and the construction of the classical stochastic integral.

Chapter 2 presents the Itô's integral (in which the integrator is the canonical Brownian motion), its definitions and properties, including both stochastic and non-stochastic properties. The chapter ends with a comparison with the version of Itô integral that was originally constructed by K. Itô, see [Itô (1944)].

Chapter 3 presents two main aspects of the Itô integral: (1) a characterisation of all Itô integrable processes based on its "belated" derivative; (2) a detailed treatment of the notion of differential that is used extensively in stochastic analysis. This is especially useful for students to have a good grasp of differentials for application purposes.

Chapter 4 gives an alternative approach of the Itô integral using the variational approach. The advantage of the variational approach is that it does not require the Vitali covering theorem. As such, we were able to consider the case of "random" intervals, that is, the interval points are random variables (or stopping time), for the case when the integrator are semi-martingales. Several versions of convergence theorems for stochastic integrals are also presented in this chapter. With the help of the (variational) convergence theorem, we showed that our definition of the integral using the generalized Riemann approach encompasses the classical stochastic integral.

In Chapters 5 and 6 we use the generalized Riemann approach to define a multiple stochastic integral with respect to Brownian motion in n-dimensions. In the definition we divide $[a, b]^n$ into two parts: (1) the diagonal part, i.e., $\{(x_1, x_2, \ldots, x_n) \in \mathbb{R}^n : x_i = x_j \text{ for some } i \neq j\}$ and (2) the non-diagonal part, i.e., $\{(x_1, x_2, \ldots, x_n) \in \mathbb{R}^n : x_i \neq x_j \text{ for all } i \neq j\}$. Note that (2) can further be divided into $n!$ contiguous sets due to the ordering of the elements $x_1, x_2, \ldots, x_n$. We establish the stochastic and non-stochastic properties of the integral, including the convergence Theorem and Fubini's Theorem. It was interesting to observe that the integral on the non-diagonal part coincides with Itô's definition of the Multiple Wiener integral. Considering both the integrals on the diagonal and the non-diagonal parts, we have proved Henstock's version of the Hu–Meyer Theorem.

Contents

Chapter 1

Introduction

In this chapter, we provide the background and the fundamentals of stochastic integrals using the generalized Riemann approach and classical stochastic integrals.

1.1 Background

It is well-known and often emphasized in texts that it is impossible to define stochastic integrals using the Riemann approach, since the integrators have paths of unbounded variation, and the integrands are highly oscillatory. The deficiency of the Riemann approach is due to the meshes, which are uniform, used in defining the Riemann sums. It is understandable that the uniform mesh is unable to handle highly oscillatory integrands and integrators. So it would be naive if "uniform meshes" is used in attempting to work out a Riemann approach to handle stochastic integrals. A way out of this apparent impasse of the Riemann approach was introduced by J. Kurzweil and R. Henstock independently in 1950s [Henstock (1955); Kurzweil (1957)]. They used non-uniform meshes (meshes that vary from point to point) instead of uniform meshes in the definition of the Riemann–Stieltjes integral. This technically minor but conceptually important modification of the classical definition of Riemann leads to the integrals which not only are more general than the Riemann–Stieltjes integral but also more general than the Lebesgue–Stieltjes integral. In the classical theory of the Riemann–Stieltjes integral, the uniform mesh size of the division is determined by a positive constant δ. Now this positive constant is replaced by a non-constant positive function $\delta(x)$, which produces non-uniform meshes of the division. Hereafter, we shall use the term "generalized Riemann approach" or "non-uniform Riemann approach" to refer to the Riemann approach where the positive constant δ is replaced by a positive function $\delta(x)$ in defining the mesh.

At first glance, this change seems to be very minor, but it turns out that it makes a profound difference in the class of functions which can be integrators or integrands. The oscillation of integrators or integrands around a point x is handled by $\delta(y)$ around x. The oscillation of continuous integrands is uniform because of the uniform continuity on a compact interval. Hence $\delta(x)$ can be chosen to be a constant. The higher the degree of the oscillation of the integrands or integrators around the point x, the smaller the value of $\delta(y)$ around x (the finer the mesh) is needed. For example, consider the function $f(x) = x^{-\frac{1}{2}}$ on $(0,1]$ and $f(0) = 0$. Certainly the oscillation of f around $x = 0$ is higher than that around $x \neq 0$. For example, given $0 < \epsilon \leq 1$, we can take $\delta(\xi) = {}^{\epsilon\xi}/_{6}$ if $\xi \neq 0$ and $\delta(0) = {}^{\epsilon^2}/_{16}$, see [Lee (1989), Example 2.7, pp. 6–7].

Therefore non-uniform meshes are needed for highly oscillating integrands. It is not surprising that this approach using non-uniform meshes has been successfully used to give alternative definitions to Itô's integral with respect to a Brownian motion, see [Boonpogkrong and Chew (2004); Chew, Tay and Toh (2001–02); Henstock (1990–91); McShane (1974, 1984); Muldowney (2012); Pop-Stojanovic (1972); Tan and Toh (2011–12); Toh (2001); Toh and Chew (2002, 2012); Xu and Lee (1992–93); Yang and Toh (2016)]. In this book, following the literature, we shall reserve the term "Itô's integral" for the classical case when the integrator is a canonical Brownian motion; if the integrator is a general martingale, local martingale or semimartingale or any other general stochastic processes, we shall just use the term "classical stochastic integral".

Throughout this book we shall let $\mathbb{R}$ denote the set of all real numbers, $\mathbb{Z}$ the set of all integers and $\mathbb{N}$ the set of all positive integers.

We shall next discuss some approaches using non-uniform meshes, which generates Riemann sums, to define stochastic integrals on compact intervals. For convenience of discussion, we shall consider the compact interval $[0,1]$.

1.1.1 *Non-uniform meshes*

Let δ be a positive function defined on $[0,1]$. An interval-point pair (I,x), where $x \in [0,1]$ and $I = (a,b]$ which is a left-open subinterval of $[0,1]$, is said to be McShane's δ-fine if $I \subset (x - \delta(x), x + \delta(x))$, i.e., the size of the mesh I depends on the associated point x, and is less than $2\delta(x)$. A finite collection D of interval-point pairs $\{(I_i, x_i)\}_{i=1}^n$ is said to be a McShane's δ-fine full division of $[0,1]$ if (i) $I_i, i = 1, 2, \ldots, n$, are disjoint left-open subintervals of $[0,1]$; (ii) $\bigcup_{i=1}^n \overline{I_i} = [0,1]$, where $\overline{I_i}$ denotes the closure of I_i, and (iii) each (I_i, x_i) is McShane δ-fine. We do not assume that $x_i \in I_i$. So x_i may be in I_i or may not be in I_i.

In the above definition, if we assume that $x \in [a, b]$, i.e., x belongs to the closure of I, then (I, x) is said to be Henstock's δ-fine or Henstock–Kurzweil's δ-fine. Accordingly, we can define a Henstock's δ-fine full division of $[0,1]$.

We remark that for any given function δ as above, McShane's and Henstock's δ-fine full divisions of $[0,1]$ exist, which are direct consequences of Heine–Borel open covering theorem [Henstock (1988), p. 33; Lee (1989), p. 4; Monteiro, Slavik and Tvrdy (2019), p. 141]. It is clear that McShane's and Henstock's δ-fine full divisions of $[0,1]$ produce non-uniform meshes which cover $[0,1]$. In the classical Riemann–Stieltjes case, a Henstock's δ-fine full division $\{(I_i, x_i)\}_{i=1}^n$ is always denoted by $\{I_i\}_{i=1}^n$ since δ is a constant positive value, the size of each I_i is always less than the constant value 2δ independent of the point x_i.

In the above, if we omit the condition (ii), then $\{(I_i, x_i)\}_{i=1}^n$ is called a McShane's (resp. Henstock's) δ-fine *partial* division of $[0,1]$.

Let $f : [0,1] \to \mathbb{R}$ and $D = \{(I_i, x_i)\}_{i=1}^n$ be a McShane's or Henstock's δ-fine full or partial division of $[0,1]$. Then the Riemann sum $S(f, \delta, D)$ of f corresponding to D is given by

$$S(f, \delta, D) = \sum_{i=1}^{n} f(x_i)|I_i|$$

where $|I_i|$ is the length of the interval I_i. As in the Riemann integral, McShane's or Henstock's integral can be defined using Riemann sums $S(f, \delta, D)$, the detail will be given in the following chapters. In this book, a division $D = \{(I_i, x_i)\}_{i=1}^n$ is often written as $D = \{(I, x)\}$ in which I represents a typical left-open subinterval of $[0,1]$ and (I, x) a typical interval-point pair. We may write

$$S(f, \delta, D) = (D) \sum f(x)|I|$$

where $D = \{(I, x)\}$.

We shall next use the above notions of non-uniform meshes and their modifications to study stochastic integrals.

1.1.1.1 *Henstock's approach using full division*

Henstock's approach to study non-stochastic integrals using non-uniform mesh is well-known, see [Henstock (1988, 1991); Kurzweil (2002)]. Henstock and Kurzweil used Henstock's δ-fine full divisions of the compact interval $[0,1]$. This method was generalized to define a stochastic integral, see [Henstock (1990–91)].

Xu J.G. and Lee P.Y., see [Xu and Lee (1992–93)], modified Henstock's approach in using non-uniform meshes to define stochastic integrals where the integrator is a Brownian motion. For each interval-point pair (I, x) which is δ-fine, the associated point x is replaced by $y(x)$, where $y(x)$ is to the left of the interval I. Accordingly, we can define a δ-fine full division. We call such a modified full-division (that is derived from Henstock's full division) the *Henstock's δ-fine belated full division.* The advantage of this modified approach is that given any $\delta(x) > 0$ on $[0,1]$, such a δ-fine belated full division of $[0,1]$ exists, and that it retains the adaptedness property of the classical stochastic integral. In fact, it was proved that the integral using this approach is equivalent to the classical Itô's integral if the integrand is classical Itô integrable. However, we observe that the construction of Henstock's δ-fine belated full division from Henstock's δ-fine full division and the proof of the uniqueness of the integral is technically involved.

1.1.1.2 *McShane's approach using full-division*

McShane's approach for non-stochastic integral is well-known and, in fact, the McShane integral is equivalent to the classical Lebesgue integral [Henstock (1988); McShane (1969)]. McShane considered McShane's δ-fine full divisions of the compact interval. Similar to Henstock's approach, this method when applied to define stochastic integral will not produce the classical stochastic integral if the integrand is non-deterministic.

Nevertheless, when the integrand is deterministic and the integrator is a Brownian motion, we shall show that the integral using McShane's approach is equivalent to the classical Itô–Wiener integral in $\mathbb{R}$ in Chapter 5. We shall further generalize this approach to compact intervals of $\mathbb{R}^m$ and prove that in fact the integral using McShane's approach is equivalent to the classical Multiple Itô–Wiener Integral. Besides the advantage of showing that the basic properties of integrals (such as additivity of integrals, integrability over sub-intervals and Cauchy's criterion) hold true, McShane's approach enables us to prove a version of Fubini's Theorem in $\mathbb{R}^m$ and subsequently the classical Hu–Meyer Theorem.

1.1.1.3 *McShane's approach using partial division*

To retain the adaptedness property of a stochastic integral on $[0,1]$ McShane considered δ-fine *partial* division for which each x of the interval-point pair (I,x) occurs to the left of I, see [McShane (1974)]. We shall call this type of division McShane's δ-fine belated partial division. More precisely, $D = \{(I_i, x_i)\}_{i=1}^n$ is a McShane's δ-fine belated partial division of $[0,1]$ if $I_i \subset (x_i, x_i + \delta(x_i))$ for each i and $I_i, i = 1, 2, \ldots, n$ are disjoint. Note that a full McShane's belated δ-fine division of $[0,1]$ may not exist, for example we may consider $\delta(x) = {}^{1-x}/_{2}$ on $[0,1]$. By using Vitali's Covering Theorem, we can only assure that a *partial* McShane's belated division that covers *almost* the entire interval $[0,1]$ exists. Hence, McShane used a partial division which covers almost $[0,1]$ to define stochastic integral on $[0,1]$.

McShane proved that this stochastic integral exists if the integrator satisfies some sort of Lipschitz conditions, see [McShane (1974)]. The Lipschitz conditions are satisfied by Brownian motions. The integrators which satisfy the Lipschitz conditions of McShane's in [McShane (1974)] are in fact semimartingales with small amendments. Furthermore, under this general setting, McShane's stochastic integral is not equivalent to the classical stochastic integral, see [Protter (1979)]. This discrepancy is due to the difference in the type of intervals used by McShane (which are right-open intervals) and the classical stochastic integrals (which use left-open intervals in defining the basic integral). In Chapter 2 of this book, by modifying the intervals as left-open intervals instead of right-open intervals into McShane's approach, we offer a direct proof of the equivalence of McShane's integral and the classical Itô integral. We shall further point out that for our stochastic case, we also have the absolute continuity property of the primitive process (it is well-known that the primitive function of a McShane integrable function in the non-stochastic case is absolutely continuous).

Next, we shall still consider δ-fine belated partial divisions, but they may not cover almost the entire $[0,1]$.

1.1.1.4 *Variational approach using belated partial division*

Motivated by classical Henstock's Lemma for the non-stochastic case, we use a variational approach to study stochastic integral. A variational approach allows us to consider *partial* divisions of $[0,1]$. For the case when the integrator is a Brownian motion, we use δ-fine belated partial divisions of $[0,1]$, which may not cover almost the entire interval, unlike McShane's approach. In Chapter 4, we prove that the integral using our variational approach is in fact equivalent to Itô's integral.

When the integrator is a generalized stochastic process, we use partial divisions consisting of "random intervals" instead of our usual intervals of $\mathbb{R}$, that is, here each I in (I, x) is a function of the probability space, which we shall denote by $I(\omega)$. Consequently, the non-uniform mesh $\delta(x)$ is replaced by a non-uniform *random* mesh $\delta(x, \omega)$. It turns out that the integral using this approach is equivalent to the classical stochastic integral if the integrator is a semimartingale.

A summary of our discussion is as shown in the table below:

Table 1.1: Summary of various generalized Riemann approaches to stochastic integrals.

Approach	**Full Div.**	**Partial Div.**	**Integrand**	**Integrator**	**Remark**
Henstock's	Yes	No	Stochastic Process	Stochastic Process	Not equivalent to classical integral, see [Henstock (1990–91)]
Henstock's belated	Yes	No	Stochastic Process	Brownian Motion	Equivalent to Itô's integral, see [Xu and Lee (1992–93)]
McShane's Full Div.	Yes	No	Deterministic Function	Brownian Motion	Equivalent to Itô–Wiener Integral, see Chapter 5
McShane's belated	No	Yes	Stochastic Process	Brownian Motion	Equivalent to Itô's Integral, see Chapter 2
Variational (usual interval)	No	Yes	Stochastic Process	Brownian Motion	Equivalent to Itô's Integral, see Chapter 4
Variational (random interval)	No	Yes	Stochastic Process	Semi-martingale	Equivalent to classical integral, see Chapter 4

1.2 Classical Approach to Stochastic Integral

The stochastic integral (see, e.g. [Chung and Williams (1990); Kopp (1985); Oksendal (1996); Protter (2005); Yeh (1995)] was first studied by K. Itô [Itô (1944)], who considered the case where the integrator is a Brownian motion, which has paths of unbounded variation. Hence, the usual notions of Riemann and Lebesgue approaches to define integrals are impossible here. The stochastic integral was first defined by Itô as follows. First, he considered the class of all "simple step" processes which are measurable and adapted to a standard filtering space (we shall define these terms in this section later). The Itô integral of this class of simple step processes is first defined as the Riemann sum with respect to a Brownian motion. It was proved that the class of all measurable simple processes is dense, under the L^2-norm, in the class of all Itô integrable processes, which are measurable and adapted to the standard filtering space. Together with Itô's isometry established and the completeness of L^2-space, the general Itô integral of such measurable adapted processes with respect to a Brownian motion is defined (the details will be outlined later in Subsection 1.2.2). This approach to define stochastic integral is further generalized to the case where the integrator is an L^2-martingale and parallel results can be established.

As mentioned in the previous section, we shall reserve the term "Itô's integral" to refer to the stochastic integral where the integrator is a Brownian motion, and where the integrator is not a Brownian motion, we just call the integral "classical stochastic integral".

In Subsection 1.2.1, we first give the general setting under which the classical stochastic integral is established. Most of the results in this section are well established, hence we shall give the references where these results can be found while the proofs of these standard results are omitted. However, the results and proofs in Subsection 1.2.2 on absolute continuity of the integral are new.

1.2.1 *Settings*

We first outline the setting under which the classical stochastic integral is established.

Let $(\Omega, \mathcal{F}, P)$ be a complete probability space and $\{\mathcal{F}_t\}$ be an increasing family of σ-subfields of $\mathcal{F}$ for $t \in [0,1]$, that is, we have $\mathcal{F}_r \subset \mathcal{F}_s$ for $0 \le r \le s \le 1$ with $\mathcal{F}_1 = \mathcal{F}$. We shall denote the probability space together with its family of increasing σ-subfields by $(\Omega, \mathcal{F}, \{\mathcal{F}_t\}, P)$, which we shall call a *standard filtering space.* We shall assume that the standard filtering space satisfies the usual conditions, that is, (1) $\mathcal{F}_0$ contains all the P-null sets; and (2) the filtration $\{\mathcal{F}_t\}$ is right-continuous, that is, $\mathcal{F}_t = \bigcap_{u>t} \mathcal{F}_u$ for all $t \in [0,1]$.

Definition 1.1. A process $X = \{X(t,\omega) : t \in [0,1]\}$ defined on a standard filtering space $(\Omega, \mathcal{F}, \{\mathcal{F}_t\}, P)$ is said to be *measurable* if it is $\mathcal{B}_{[0,1]} \times \mathcal{F}$-measurable, where $\mathcal{B}_A$ denotes the Borel σ-field on the set A.

Notation 1.1. In this book, we denote $X(t,\omega)$ by $X_t(\omega)$. Also, we use the notation X_t when we want to highlight the function X_t defined on all $\omega \in \Omega$ given by

$$X_t : \omega \to X(t,\omega)$$

for each $t \in [0,1]$.

Definition 1.2. A measurable process $X = \{X(t,\omega) : t \in [0,1]\}$ defined on a standard filtering space $(\Omega, \mathcal{F}, \{\mathcal{F}_t\}, P)$ is said to be *adapted to the filtration* $\{\mathcal{F}_t\}$ if X_t is $\mathcal{F}_t$-measurable for each $t \in [0,1]$. Given a measurable process, it follows from definition that the smallest filtration $\{\mathcal{F}_t\}$ can be constructed by choosing $\mathcal{F}_t = \sigma\{X(s,\omega) : s \leq t\}$, where σ denotes the smallest σ-field containing $\{X(s,\omega) : s \leq t\}$. So if $\{\mathcal{G}_t\}$ is any other set of filtration of $\mathcal{F}$ such that X is adapted, then $\mathcal{F}_t \subset \mathcal{G}_t$ for each $t \in [0,1]$.

Definition 1.3. A measurable process $X = \{X(t,\omega) : t \in [0,1]\}$ defined on $(\Omega, \mathcal{F}, \{\mathcal{F}_t\}, P)$ is said to be a *martingale* if

(i) X is adapted to the filtration, that is, X_t is $\mathcal{F}_t$-measurable for each $t \in [0,1]$;
(ii) $\int_\Omega |X_t| dP < \infty$ for almost all $t \in [0,1]$; and
(iii) $E(X_t \mid \mathcal{F}_s) = X_s$ for all $t \geq s$, where $E(X_t|\mathcal{F}_s)$ is the conditional expectation of X_t given $\mathcal{F}_s$, which is defined to be a random variable such that $E(X_t \mid \mathcal{F}_s)$ is $\mathcal{F}_s$-measurable and

$$\int_A E(X_t \mid \mathcal{F}_s) dP = \int_A X_t dP$$

for each $A \in \mathcal{F}_s$. By the Radon-Nikodym Theorem, $E(X_t \mid \mathcal{F}_s)$ is well-defined.

Remark 1.1. We make the following observations:

(i) It is easy to see that

$$E(X_s) = E\left(E(X_t \mid \mathcal{F}_s)\right) = E(X_t)$$

for all $t \geq s$, i.e., $E(X_t)$ is constant for all $t \in [0,1]$.
(ii) The difference of two martingales is again a martingale.

Let $L^2(\Omega)$ or L^2 be the space of all measurable functions f such that $\int_\Omega |f|^2 dP < \infty$.

A martingale X is said to be an L^2-martingale if in addition to conditions (i), (ii) and (iii) in Definition 1.3, we have $\sup_{t\in[0,1]} \int_\Omega |X_t|^2 dP$ is finite.

Remark 1.2. We make the following observations:

(i) Let X be an L^2-martingale. Then X has the orthogonal increment property

$$E[(X_v - X_u)(X_t - X_s)] = 0$$

for $0 \leq u < v < s < t \leq 1$. This can be seen as follows:

$$\begin{aligned} E[(X_t - X_s)(X_v - X_u)] &= E\{E[(X_t - X_s)(X_v - X_u) \mid \mathcal{F}_s]\} \\ &= E\{(X_v - X_u)E[(X_t - X_s) \mid \mathcal{F}_s]\} \\ &= 0. \end{aligned}$$

The above step follows from the definition of a martingale that $E[X_t - X_s|\mathcal{F}_s] = 0$ whenever $s < t$.

With the above orthogonal property, we get

$$E\left|\sum[X_v - X_u]\right|^2 = E\left(\sum[X_v - X_u]^2\right)$$

for any finite collection of disjoint intervals $\{(u, v]\}$, since

$$\begin{aligned} &E\left|\sum[X_v - X_u]\right|^2 \\ &= E\left\{\sum_i (X_{v_i} - X_{u_i})^2 + 2\sum_{i<j}(X_{v_i} - X_{u_i})(X_{v_j} - X_{u_j})\right\} \\ &= \sum E(X_v - X_u)^2 + 2\sum_{i<j} E(X_{v_i} - X_{u_i})(X_{v_j} - X_{u_j}) \\ &= \sum E(X_v - X_u)^2 + 0. \end{aligned}$$

(ii) It is also clear that for $u < v$ we have

$$E[X_v X_u] = E\left[E[X_v X_u \mid \mathcal{F}_u]\right] = E\left[X_u E(X_v \mid \mathcal{F}_u)\right] = E(X_u^2)$$

so that we have

$$E\left[X_v - X_u\right]^2 = E[X_v^2 - X_u^2].$$

Without loss of generality, we may assume that $X(0, \omega) = 0$ for all $\omega \in \Omega$. Also, for succinctness, we will denote $\int_\Omega f(\omega)dP$ by $E(f)$, where we fix the complete probability space $(\Omega, \mathcal{F}, P)$.

Definition 1.4.

(i) A measurable process $X = \{X(t, \omega) : t \in [0, 1]\}$ is said to be *continuous* if the mapping $t \to X(t, \omega)$ is continuous for each $\omega \in \Omega$. In other words, continuity is understood as sample path continuity.

(ii) A measurable process X is said to be a.s. continuous on $[0,1]$ if for almost all $\omega \in \Omega$, the function $X(\cdot,\omega)$ is continuous on $[0,1]$.
(iii) A measurable process is said to be *cadlag* if for almost all $\omega \in \Omega$, the sample paths $X_t : t \to X(t,\omega)$ are right-continuous with left limits.

Throughout this book, all measurable processes functioning as the integrators in all our subsequent chapters that we consider are assumed to be *cadlag*.

Definition 1.5. Let $\mathcal{G}$ be the smallest σ-algebra of $[0,1] \times \Omega$ such that all left-continuous $\{\mathcal{F}_t\}$-adapted processes are $\mathcal{G}$-measurable. Then $\mathcal{G}$ is a $\{\mathcal{F}_t\}$-predictable σ-algebra. A stochastic process is said to be $\{\mathcal{F}_t\}$-predictable if it is $\mathcal{G}$-measurable.

It was proved in [Yeh (1995), Lemma 2.20, p. 21] that a $\{\mathcal{F}_t\}$-predictable process is adapted to the filtration $\{\mathcal{F}_t\}$.

Definition 1.6. A measurable process $B = \{B_t(\omega) : t \in [0,1]\}$ defined on a standard filtering space $(\Omega, \mathcal{F}, \{\mathcal{F}_t\}, P)$ is said to be a canonical *Brownian motion* (or *Wiener Process*) if it has the following properties:

(i) $B_0(\omega) = 0$ for all $\omega \in \Omega$;
(ii) **(Normal increments)** $B_t - B_s$ has a Normal distribution with mean 0 and variance $t - s$ for all $t > s$. This implies that B_t has a Normal distribution with mean 0 and variance t;
(iii) **(Independence of Increments)** $B_t - B_s$ is independent of the past, that is, $B_u, 0 \le u < s < t$; and
(iv) **(Continuity of Paths)** B is continuous (see Definition 1.4(i)).

It is well-known that a canonical Brownian motion is in fact a continuous martingale, see [Calin (2015); Klebaner (2012); Mikosch (1998)].

Definition 1.7. For any L^2-martingale X (for example, a Brownian motion), there exists a unique predictable increasing process $\langle X \rangle$ with $\langle X \rangle_0 = 0$ such that $X^2 - \langle X \rangle$ is also an L^2-martingale, for example, see [Durrett (1990), pp. 52–55; Yeh (1995), Proposition 11.20, pp. 212–213]. We shall call $\langle X \rangle$ the *quadratic variation process* associated with X.

Continuity of X necessarily implies the continuity of $\langle X \rangle$, see [Ikeda and Watanabe (1981), Section 2, p. 53]. In general, the quadratic variation process of a *cadlag* process is *cadlag*, see [Protter (2005), Theorem 22, p. 59]. If X is a Brownian motion, then it is well-known that $\langle X \rangle_t = t$, which is deterministic.

1.2.2 *Construction of classical stochastic integral*

1.2.2.1 L^2-*martingale as integrator*

First, consider the case when the integrator is an L^2-martingale. Let $\mathcal{L}_2^{\langle X\rangle}$ denote the class of all measurable processes $\varphi = \{\varphi(t,\omega) : t \in [0,1]\}$ adapted to a standard filtering space $(\Omega, \mathcal{F}, \{\mathcal{F}_t\}, P)$ and $\{\mathcal{F}_t\}$-predictable such that

$$\|\varphi\|_{\mathcal{L}^2}^2 = \int_\Omega \int_0^1 \varphi^2(t,\omega) d\langle X\rangle_t dP < \infty.$$

It is easy to verify that $\|\cdot\|_{\mathcal{L}^2}$ is a norm on $\mathcal{L}_2^{\langle X\rangle}$.

When $X = B$ is a canonical Brownian motion where $\langle B\rangle_t \equiv t$ for all $t \in [0,1]$, we shall use the notation $\mathcal{L}_2$ to denote the class of all measurable processes $\varphi = \{\varphi(t,\omega) : t \in [0,1]\}$ adapted to the standard filtering space $(\Omega, \mathcal{F}, \{\mathcal{F}_t\}, P)$ such that

$$\|\varphi\|_{\mathcal{L}_2}^2 = \int_\Omega \int_0^1 \varphi^2(t,\omega) dt dP < \infty.$$

When $X = B$, we may ignore the predictability of the integrand φ.

Let $\mathcal{L}_0$ denote the class of all bounded adapted left-continuous (hence predictable) simple processes on $[0,1]$, that is, every $\varphi \in \mathcal{L}_0$ can be written in the form

$$\varphi(t,\omega) = \xi_0(\omega) 1_{\{0\}}(t) + \sum_{k=1}^n \xi_{k-1}(\omega) 1_{(t_{k-1},t_k]}(t), \tag{1.1}$$

where $0 = t_0 < t_1 < t_2 < \cdots < t_{n-1} < t_n = 1$ and each ξ_k is $\mathcal{F}_{t_k}$-measurable (so it is clear that φ is adapted and measurable), and there exists $M > 0$ such that $|\xi_k(\omega)| \le M$ for all $\omega \in \Omega$ and $k = 1, 2, \ldots, n$.

For each $\varphi \in \mathcal{L}_0$ of the form (1.1), the classical stochastic integral with respect to the martingale X is defined as

$$(I)\int_0^1 \varphi(t,\omega) dX(t,\omega) = \sum_{k=1}^n \xi_{k-1}(\omega)[X(t_k,\omega) - X(t_{k-1},\omega)]$$

for each $\omega \in \Omega$; or perceiving it as an operator on Ω we may write

$$(I)\int_0^1 \varphi_t dX_t = \sum_{k=1}^n \xi_{k-1}[X_{t_k} - X_{t_{k-1}}].$$

We shall use the symbol $(I)\int$ to denote the classical stochastic integral throughout this book.

Lemma 1.1. *[Chung and Williams (1990), p. 37; Yeh (1995), Proposition 12.8, p. 235] For any $\varphi \in \mathcal{L}_2^{\langle X\rangle}$, there exists a sequence of simple processes in $\mathcal{L}_0$ which converges to φ under the norm of $\|\cdot\|_{\mathcal{L}^2}$; in other words, the class $\mathcal{L}_0$ of simple processes is dense in $\mathcal{L}_2^{\langle X\rangle}$ under $\|\cdot\|_{\mathcal{L}_2}$.*

Lemma 1.2 (Isometric Equality). *[Chung and Williams (1990), p. 37] For every $\varphi \in \mathcal{L}_0$, we have*

$$E\left|(I)\int \varphi_t dX_t\right|^2 = \|\varphi\|_{\mathcal{L}^2}^2 = \int_\Omega \int_0^1 \varphi^2(t,\omega)d\langle X\rangle_t dP.$$

In the case when $X = B$ is a canonical Brownian motion, we have

$$E\left((I)\int_0^1 \varphi_t dB_t\right)^2 = E\left[\int_0^1 \varphi^2(t,\omega)dt\right] = \int_0^1 E[\varphi_t^2]dt, \tag{1.2}$$

which is called Itô's isometry.

With the above two lemmas, the classical stochastic integral for a general $\varphi \in \mathcal{L}_2^{\langle X\rangle}$ is defined as follows:

Let $\varphi \in \mathcal{L}_2^{\langle X\rangle}$. Then there exists a sequence $\{\varphi^n\} \subset \mathcal{L}_0$, which converges to φ in $\|\cdot\|_{\mathcal{L}^2}$-norm. Completeness of $L_2(\Omega)$ together with Isometric Equality (see Lemma 1.2) implies that

$$L_2-\lim_{n\to\infty}(I)\int_0^1 \varphi^n(t,\cdot)dX(t,\cdot)$$

exists. We define

$$(I)\int_0^1 \varphi_t dX_t = L_2-\lim_{n\to\infty}(I)\int_0^1 \varphi_t^n dX_t,$$

where the limit is taken in L^2-norm. We have the following elementary result:

Theorem 1.1. *[Oksendal (1996), Theorem 3.7, p. 22]*

(i) *If the measurable adapted process φ is classical stochastic integrable with respect to X on $[0,1]$, then it is classical stochastic integrable on every $[a,b] \subset [0,1]$.*

(ii) *If the measurable adapted process φ is classical stochastic integrable with respect to X on $[a,b]$ and $[b,c] \subset [0,1]$, then it is classical stochastic integrable on $[a,c]$ and*

$$(I)\int_a^c \varphi_t dX_t = (I)\int_a^b \varphi_t dX_t + (I)\int_b^c \varphi_t dX_t.$$

We remark that the proofs of Theorem 1.1 can be established by first proving the results for simple adapted measurable processes and then taking limits.

In the following we shall establish some results on the absolute continuity of the integral, which are not discussed in the classical literature on stochastic integrals.

1.2.2.2 *Absolute continuity of primitive process*

It is well-known from the classical theory of integration that if f is Lebesgue integrable on $[0,1]$, then its primitive function F defined by $F_t = \int_0^t f(s)ds$ is absolutely continuous on $[0,1]$, see for example [Royden (1989), Theorem 14, p. 110]. Next we shall attempt to establish results that the primitive process Φ of a classical Itô integrable process also satisfies some form of absolute continuity condition.

We remark that if the integrator X is a Brownian motion, then from Itô's isometry (1.2), we can get the absolute continuity easily. However, when the integrator X is not a Brownian motion, it is not clear that we can get results of absolute continuity from Lemma 1.2, since Fubini's Theorem cannot be applied. Note that $\langle X \rangle$ is a function defined on $[0,1] \times \Omega$ and P depends on Ω; the two integrators cannot be interchanged as in the situation of (1.2).

Definition 1.8. A stochastic process $Y = \{Y(t,\omega) : t \in [0,1]\}$ is said to be $AC^2[0,1]$ if given $\epsilon > 0$ there exists $\eta > 0$ such that

$$E\left|\sum [Y_v - Y_u]\right|^2 < \epsilon$$

for any finite collection of disjoint intervals $\{(u,v]\}$ for which $\sum |v-u| < \eta$.

Definition 1.9. Given the stochastic processes $Y = \{Y(t,\omega) : t \in [0,1]\}$ and $X = \{X(t,\omega) : t \in [0,1]\}$, the process Y is said to be ΔX-$AC^2[0,1]$ if given $\epsilon > 0$ there exists $\eta > 0$ such that

$$E\left|\sum [Y_v - Y_u]\right|^2 < \epsilon$$

for any finite collection of disjoint intervals $\{(u,v]\}$ of $[0,1]$ for which

$$E\left|\sum [X_v - X_u]\right|^2 < \eta.$$

Theorem 1.2. *Let X be a continuous L^2-martingale with its quadratic variation $\langle X \rangle$. Assume that $E\langle X \rangle_t$ is absolutely continuous on $[0,1]$. A stochastic process Y is $AC^2[0,1]$ if Y is ΔX-$AC^2[0,1]$. In particular, the result is true if X is a canonical Brownian motion.*

Proof. We just need to show that we have

$$E\left|\sum[X_v - X_u]\right|^2 = \sum\{E\langle X\rangle_v - E\langle X\rangle_u\} \tag{1.3}$$

for any finite collection of disjoint intervals $\{(u, v]\}$. From the orthogonal increment property of martingales, we have

$$E\left|\sum[X_v - X_u]\right|^2 = \sum E(X_v - X_u)^2, \tag{1.4}$$

see Remark 1.2(i) after Definition 1.3. It is easy to see that

$$E(X_v - X_u)^2 = E(X_v^2 - X_u^2)$$

for any $u < v$ and further from the remark after Definition 1.3 and the fact that $X^2 - \langle X\rangle$ is a martingale, for any $u < v$, we have

$$E\left(X_v^2 - \langle X\rangle_v\right) = E\left(X_u^2 - \langle X\rangle_u\right)$$

so that we have

$$E\left(X_v^2 - X_u^2\right) = E\left(\langle X\rangle_v - \langle X\rangle_u\right),$$

thereby completing our proof. □

We remark that the next few results are crucial for our subsequent development of stochastic integrals.

Lemma 1.3 (Variational Convergence). *Let X be an L^2-martingale. Suppose that $(I)\int_0^1 \varphi_t^n dX_t \to (I)\int_0^1 \varphi_t dX_t$ in $L_2(\Omega)$ as $n \to \infty$. Then given $\epsilon > 0$ there exists a positive integer N such that whenever $\{(u, v]\}$ is a finite collection of disjoint subintervals of $[0, 1]$, we have*

$$E\left|\sum(I)\int_u^v \varphi_t^n dX_t - \sum(I)\int_u^v \varphi_t dX_t\right|^2 < \epsilon$$

for any $n \geq N$.

Proof. Let $\epsilon > 0$ be given. If $\varphi \in \mathcal{L}_2^{\langle X\rangle}$, let $\Phi_t = (I)\int_0^t \varphi_s dX_s$ be the primitive process. Then Φ_t is an L^2-martingale, see [Yeh (1995), Remark 12.15, p. 241]. Similarly, for each $n \in \mathbb{N}$, if $\Phi_t^n = (I)\int_0^t \varphi_s^n dX_s$, then each Φ_t^n is also an L^2-martingale. Choose N such that

$$E\left|(I)\int_0^1 \varphi_t^n dX_t - (I)\int_0^1 \varphi_t dX_t\right|^2 < \epsilon$$

whenever $n \geq N$.

Given a finite collection of disjoint subintervals $D = \{(u, v]\}$ from $[0, 1]$, let $(D)\sum$ denote the summation over those intervals included in D and

$(D^c)\sum$ be the summation over left-open subintervals of $(0,1]$ not included in D. Then $D \cup D^c = (0,1]$ and

$$(D \cup D^c)\sum := (D)\sum + (D^c)\sum.$$

Thus

$$\begin{aligned}
E&\left|(D)\sum(I)\int_u^v \varphi_t^n dX_t - (D)\sum(I)\int_u^v \varphi_t dX_t\right|^2 \\
&= E\left|(D)\sum(\Phi_v^n - \Phi_v^n) - (D)\sum(\Phi_v - \Phi_u)\right|^2 \\
&= E\left|(D)\sum\{(\Phi_v^n - \Phi_v) - (\Phi_u^n - \Phi_u)\}\right|^2 \\
&= E\left|(D)\sum(\Psi_v^n - \Psi_u^n)\right|^2,
\end{aligned}$$

where $\Psi_t^n = \Phi_t^n - \Phi_t$, which is also an L^2-martingale since the difference of two martingales is again a martingale (see Remark 1.1(ii) after Definition 1.3). From the orthogonal increment of martingales, we thus have

$$\begin{aligned}
E\left|(D)\sum(\Psi_v^n - \Psi_u^n)\right|^2 &= (D)\sum E(\Psi_v^n - \Psi_u^n)^2 \\
&\leq (D)\sum E(\Psi_v^n - \Psi_u^n)^2 + (D^c)\sum E(\Psi_v^n - \Psi_u^n)^2 \\
&= E\left|(D \cup D^c)\sum(\Psi_v^n - \Psi_u^n)\right|^2 \\
&= E\left|\Psi_1^n - \Psi_0^n\right|^2 \qquad (1.5) \\
&= E\left|(I)\int_0^1 \varphi_t^n dX_t - (I)\int_0^1 \varphi_t dX_t\right|^2 \\
&< \epsilon,
\end{aligned}$$

we remark that (1.5) is a consequence of (1.4), thereby completing the proof. □

From the above proof, it is clear that we have

Theorem 1.3. *Let $\{X^n\}$ be a sequence of L^2-martingales such that $X_1^n - X_0^n \to X_1 - X_0$ in $L^2(\Omega)$, where X is an L^2-martingale. Then for any $\epsilon > 0$, there exists a positive integer N such that for any finite collection of disjoint subintervals $\{(u,v]\}$ of $[0,1]$, we have*

$$E\left|\sum(X_v^n - X_u^n) - (X_v - X_u)\right|^2 < \epsilon,$$

whenever $n \geq N$.

Theorem 1.4. *Let X be an L_2-martingale with $\langle X \rangle$ the corresponding quadratic variation process, and let $\varphi = \{\varphi(t, \omega) : t \in [0, 1]\}$ be an adapted predictable measurable process from $\mathcal{L}_2^{\langle X \rangle}$ (which is hence classical stochastic integrable). Define $\Phi_t = (I) \int_0^t \varphi_s dX_s$ for all $t \in [0, 1]$. Then Φ is ΔX-$AC^2[0, 1]$.*

Proof. Let $\epsilon > 0$ be given.

Step 1

First, we shall prove the result for the case of measurable adapted processes in $\mathcal{L}_0$, which are bounded and left-continuous.

Proof of Step 1

Consider a simple measurable adapted process $\varphi = \{\varphi(t, \cdot) : t \in [0, 1]\}$ in $\mathcal{L}_0$ with its associated sequence $0 = t_0 < t_1 < t_2 < \cdots < t_n = 1$ and let

$$\varphi(t, \omega) = \begin{cases} \xi_{k-1}(\omega), & \text{if } t \in (t_{k-1}, t_k], k = 1, 2, \ldots, n \\ \xi_0(\omega), & \text{if } t = 0 \end{cases}$$

and let $M > 0$ be such that $|\xi_k(\omega)| \leq M$ for all $k = 0, 1, 2, \ldots, n$, and each ξ_k is $\mathcal{F}_{t_k}$-measurable. We first show that the primitive process $\Phi_t = (I) \int_0^t \varphi_s dX_s$ is ΔX-$AC^2[0, 1]$.

Let $\eta = \frac{\epsilon}{M^2}$ and consider any finite collection of disjoint intervals $\{(u, v]\}$ with

$$E \left| \sum [X_v - X_u] \right|^2 < \eta.$$

We have

$$\begin{aligned} E \left| \sum [\Phi_v - \Phi_u] \right|^2 &= E \left| \sum \xi_i(\cdot)[X_v - X_u] \right|^2 \\ &\leq M^2 E \left| \sum [X_v - X_u] \right|^2 \\ &\leq \epsilon, \end{aligned}$$

thereby showing that Φ is ΔX-$AC^2[0, 1]$.

Step 2

Let $\varphi \in \mathcal{L}_2^{\langle X \rangle}$ and $\{\varphi^n\}$ be a sequence in $\mathcal{L}_0$ such that $\varphi^n \to \varphi$ in $\mathcal{L}_2^{\langle X \rangle}$-norm.

From Lemma 1.3, choose a positive integer N such that for any collection of disjoint intervals $\{(u, v]\}$ we have, when $n \geq N$,

$$E \left| \sum (I) \int_u^v \varphi_t^n dX_t - \sum (I) \int_u^v \varphi_t dX_t \right|^2 < \frac{\epsilon}{4}.$$

The choice of N is independent of the collection of intervals by Lemma 1.3. Choose an $n \geq N$ and fix such a value of n. Then by Step 1 above, φ^n is ΔX-$AC^2[0,1]$ for each $n \in \mathbb{N}$ since it is a process in $\mathcal{L}_0$. Choose $\eta_1 > 0$ such that

$$E\left|\sum(I)\int_u^v \varphi_t^n dX_t\right|^2 < \frac{\epsilon}{4}$$

whenever

$$E\left|\sum(X_v - X_u)\right|^2 < \eta_1.$$

Hence

$$\begin{aligned} E\left|\sum(I)\int_u^v \varphi_t dX_t\right|^2 &= E\left|\sum(I)\int_u^v (\varphi_t - \varphi_t^n)dX_t + \sum(I)\int_u^v \varphi_t^n dX_t\right|^2 \\ &\leq 2E\left|\sum(I)\int_u^v (\varphi_t^n - \varphi_t)dX_t\right|^2 + 2E\left|\sum(I)\int_u^v \varphi_t^n dX_t\right|^2 \\ &\leq \epsilon \end{aligned}$$

whenever we have a finite collection of disjoint interval $\{(u,v]\}$ with

$$E\left|\sum(X_v - X_u)\right|^2 < \eta_1,$$

thereby completing our proof. □

1.2.2.3 *Local martingale and semimartingale as integrators*

The classical stochastic integral with respect to a local martingale is defined in the following section. First, we state some standard definitions before defining the integral.

Definition 1.10. [Yeh (1995), Definition 3.1, p. 25] Let $T : \Omega \to [0,\infty]$ be a random variable defined on $(\Omega, \mathcal{F}, \{\mathcal{F}_t\}, P)$. Then T is a *stopping time* if

$$\{\omega \in \Omega : T(\omega) \leq t\} \in \mathcal{F}_t$$

for each $t \geq 0$.

Definition 1.11. [Yeh (1995), Definition 14.1, p. 298] Let $X = \{X_t : t \in [0,1]\}$ and let $\{S_n : n \in \mathbb{N}\}$ be an increasing sequence of stopping times such that $S_n(\omega) \to 1$ as $n \to \infty$ for almost all $\omega \in \Omega$. Then X is said to be a *local martingale* if $X^{S_n \wedge}$ is a martingale for every $n \in \mathbb{N}$, where $X^{S_n \wedge}(t,\omega) = X(S_n(\omega) \wedge t, \omega)$. For any $a, b \in \mathbb{R}$, let $a \wedge b = \min\{a,b\}$. Then X is said to be an *L^2-local martingale* if each $X^{S_n \wedge}$ is an L^2-martingale for every $n \in \mathbb{N}$.

The classical stochastic integral of a stochastic process φ with respect to a local martingale X is defined as follows. The details of the construction can be referred to [Yeh (1995), Section 14, pp. 306–317].

There exist a sequence $\{\varphi^n\}$ such that $\varphi^n(t,\omega) \to \varphi(t,\omega)$ as $n \to \infty$ for almost all $\omega \in \Omega$ and all $t \in [0,1]$, and a sequence $\{X^{S_n\wedge}\}$ of L^2-martingales such that φ^n is classical stochastic integrable with respect to each $X^{S_k\wedge}$ for each $n, k \in \mathbb{N}$. Then the stochastic integral $(I)\int_0^1 \varphi_t^n dX_t^{S_n\wedge}$ is defined as in the earlier section, since each $X^{S_n\wedge}$ is a martingale, $n = 1, 2, \ldots$. Then the classical stochastic integral of φ with respect to X is defined as the pointwise limit

$$(I)\int_0^1 \varphi_t dX_t = \lim_{n\to\infty} (I)\int_0^1 \varphi^n dX_t^{S_n\wedge}.$$

Next we quote another useful result which we shall need in Chapter 4, Subsection 4.3.2 to prove our equivalence theorem later.

Lemma 1.4. *[Yeh (1995), Theorem 14.16, p. 309] Let φ be classical stochastic integrable with respect to the L^2 local-martingale X with the associated sequence $\{S_n\}$ (see Definition 1.11). Then φ is classical stochastic integrable with respect to each $X^{S_n\wedge}$ for each $n \in \mathbb{N}$ and*

$$(I)\int_0^1 \varphi_t dX_t = \lim_{n\to\infty} (I)\int_0^1 \varphi_t dX_t^{S_n\wedge}$$

for almost all $\omega \in \Omega$.

Next we consider the integrator $X = \{X_t : t \in [0,1]\}$ to be a semimartingale. Then we may write $X = L + V$, where L is an L^2-local martingale, and V is a stochastic process with bounded variation. Hence the classical stochastic integral of φ with respect to X on $[0,1]$ is defined as

$$(I)\int_0^1 \varphi_t dX_t = (I)\int_0^1 \varphi_t dL_t + \int_0^1 \varphi_t dV_t$$

where the first integral on the right is the classical stochastic integral of φ with respect to the local martingale L and the second integral is the usual pathwise Lebesgue–Stieltjes integral.

Chapter 2

The Itô Integral

In this chapter, we shall use the non-uniform (generalized) Riemann approach to define the Itô integral. The Riemann sum is given by belated δ-fine partial division. The belated δ-fine full division may not exist, therefore the Vitali cover needed in the definition. The necessary and sufficient condition is given for an adapted Stochastic process to be Itô-integrable. Itô's isometry and Itô's Lemma are proved using Henstock's Lemma and absolute continuity. An alternative definition of the Itô integral using a variation approach is given. The Vitali cover is not needed in the variation approach. Finally, the Itô integral and the classical Itô integral are proved to be equivalent.

2.1 Definition of the Integral

Let B be the canonical Brownian motion adapted to the standard filtering space $(\Omega, \mathcal{F}, \{\mathcal{F}_t\}, P)$ as discussed in the previous chapter. Let f be a real-valued stochastic process, that is, $f : [a, b] \times \Omega \to \mathbb{R}$, such that for any $t \in [a, b]$, $f_t(\cdot) = f(t, \cdot)$ is a random variable on Ω. Furthermore, in this chapter, we always assume that f is adapted to the standard filtering space $(\Omega, \mathcal{F}, \{\mathcal{F}_t\}, P)$ and $[a, b] \subseteq [0, \infty)$.

Definition 2.1. A finite collection of left-open interval $\{(u_i, v_i]\}_{i=1}^n$ is said to be a *partial partition* of $[a, b]$ if

(i) the intervals $(u_i, v_i]$ are disjoint, that is, for any $i \neq j$, $(u_i, v_i] \cap (u_j, v_j]$ is empty; and

(ii) $\bigcup_{i=1}^n (u_i, v_i] \subset [a, b]$.

If $\bigcup_{i=1}^n (u_i, v_i] = (a, b]$, then $\{(u_i, v_i]\}_{i=1}^n$ is said to be a *full partition* of $(a, b]$.

Definition 2.2. Let δ be a positive function on $[a, b]$. A finite collection D of interval-point pairs $\{((\xi_i, v_i], \xi_i)\}_{i=1}^n$ is a δ-fine *belated* partial division of $[a, b]$ if

(i) $\{(\xi_i, v_i]\}_{i=1}^n$ is a partial partition of $[a, b]$; and
(ii) each $(\xi_i, v_i]$ is δ-fine belated, that is, $(\xi_i, v_i] \subset (\xi_i, \xi_i + \delta(\xi_i))$.

For brevity, a δ-fine belated partial division D in Definition 2.2 is often written as $D = \{((\xi, v], \xi)\}$.

Notice that the term *belated* is used in Definition 2.2 since the tag ξ_i to each interval $(\xi_i, v_i]$ is the left-hand point of the interval.

Definition 2.3. Given $\eta > 0$, the δ-fine belated partial division in Definition 2.2, $D = \{((\xi_i, v_i], \xi_i)\}_{i=1}^n$ is said to be (δ, η)-fine if the intervals $\{(\xi_i, v_i]\}_{i=1}^n$ fail to cover $[a, b]$ by at most Lebesgue measure η, that is, if

$$\left| b - a - \sum_i (v_i - \xi_i) \right| < \eta.$$

In our definition above, belated partial divisions are used. There are two questions arising from this:

(1) Given a positive function $\delta > 0$ on $[a, b]$, is it always possible to find a *full* δ-fine belated division that covers the left-open interval $(a, b]$?
(2) Given any $\eta > 0$ and a positive function $\delta > 0$ on $[a, b]$, is it always possible to find a (δ, η)-fine belated partial division of $[a, b]$?

In the following part of this section, the two questions above are addressed.

To answer the first question: given a positive function $\delta > 0$ on $[a, b]$, we may not be able to find a *full* δ-fine belated division that covers the entire interval $(a, b]$. For example, let $\delta(\xi) = (b - \xi)/2$ for all $\xi \in [a, b)$. Take any δ-fine belated (partial or full) division $D = \{((\xi_i, v_i], \xi_i)\}_{i=1}^n$. Assume the order

$$\xi_1 < v_1 \leq \xi_2 < v_2 \leq \xi_3 < v_3 \leq \cdots \leq \xi_n < v_n.$$

Note that $(\xi_n, v_n] \subset (\xi_n, \xi_n + \delta(\xi_n))$ and $\xi_n + \delta(\xi_n) = \xi_n + \frac{b-\xi_n}{2} = \frac{\xi_n+b}{2} < \frac{b+b}{2} = b$. Thus $(\xi_n, v_n] \subset (\xi_n, b)$. Hence $(v_n, b]$ is not covered by intervals in D. Therefore δ-fine belated full division may not exist.

To answer the second question, we need the Vitali's Covering Theorem (Theorem 2.1), see e.g. Chapter 5 in [Royden (1989)].

Definition 2.4. Let G be a subset of $\mathbb{R}$. A collection $\mathcal{I}$ of intervals is a Vitali cover of G if for each $\epsilon > 0$ and any $x \in G$, there exists $I \in \mathcal{I}$ such that $x \in I$ and $|I| < \epsilon$, where $|I|$ is the length of I.

Theorem 2.1. *Let G be a bounded subset of $\mathbb{R}$ and $\mathcal{I}$ be a Vitali cover of G. Then, given $\epsilon > 0$, there is a finite disjoint collection $\{I_1, I_2, \ldots, I_n\}$ of intervals in $\mathcal{I}$ such that*

$$\mu\left[G \setminus \bigcup_{i=1}^{n} I_i\right] < \epsilon,$$

where μ is Lebesgue outer measure. Suppose $G = [a, b]$. Then $G \setminus \bigcup_{i=1}^{n} I_i$ is a finite union of disjoint intervals, say $\bigcup_{j=1}^{m} J_j$. Hence $\mu\left[G \setminus \bigcup_{i=1}^{n} I_i\right] = \sum_{j=1}^{m} |J_j|$.

Lemma 2.1. *Given a positive function $\delta > 0$ on $[a, b]$, and any positive number $\eta > 0$, it is always possible to find a (δ, η)-fine belated partial division of $[a, b]$.*

Proof. Let $D = \{((\xi, v], \xi)\}$ be δ-fine belated partial division and $P(D) = \{[\xi, v]\}$ be the corresponding partial partition. Let $\mathcal{I} = \bigcup\{P(D) :$ all δ-fine belated partial divisions D of $[a, b]\}$. Note that we use $[\xi, v]$ instead of $(\xi, v]$, since $\xi \in [\xi, v]$ and $|[\xi, v]| = |(\xi, v]|$. Then $\mathcal{I}$ is a Vitali cover of $[a, b]$. By Vitali's Covering Theorem (Theorem 2.1), a (δ, η)-fine belated partial division of $[a, b]$ exists. □

We are now ready to give a definition of the Itô stochastic integral.

Definition 2.5. Let $f : [a, b] \times \Omega \to \mathbb{R}$ be an adapted stochastic process. Then f is said to be Itô integrable on $[a, b]$ if there exists an $A \in L^2(\Omega)$ such that for any $\epsilon > 0$, there exist a positive function δ on $[a, b]$ and a positive number $\eta > 0$ such that for any (δ, η)-fine belated partial division $D = \{((\xi_i, v_i], \xi_i)\}_{i=1}^{n}$ of $[a, b]$, we have

$$E\left(\sum_{i=1}^{n} f_{\xi_i}(B_{v_i} - B_{\xi_i}) - A\right)^2 < \epsilon.$$

Example 2.1. Let $c \in \mathbb{R}$ and $f(t, \omega) = c$ for all $t \in [a, b]$ and all $\omega \in \Omega$. Then f is Itô integrable on $[a, b]$ with the value $c(B_b - B_a) \in L^2(\Omega)$.

Proof. We shall assume that $c \neq 0$ since the case of $c = 0$ is obvious. Let $\epsilon > 0$. Choose any positive function δ on $[a, b]$ and $\eta = \epsilon/c^2$. Let $D = \{((\xi_i, v_i], \xi_i)\}_{i=1}^{n}$ be any (δ, η)-fine belated partial division of $[a, b]$. Then

$(a,b] \setminus \bigcup_{i=1}^{n}(\xi_i, v_i]$ is the finite union of disjoint intervals, say $\bigcup_{j=1}^{m}(a_j, b_j]$ and $\sum_{j=1}^{m}|b_j - a_j| < \eta$. Hence

$$\begin{aligned} E\left(\sum_{i=1}^{n} f_{\xi_i}(B_{v_i} - B_{\xi_i}) - c(B_b - B_a)\right)^2 &= E\left(\sum_{j=1}^{m} c(B_{b_j} - B_{a_j})\right)^2 \\ &= c^2 E\left(\sum_{j=1}^{m}(B_{b_j} - B_{a_j})\right)^2 \\ &= c^2 \sum_{j=1}^{m} E\left(B_{b_j} - B_{a_j}\right)^2 \\ &= c^2 \sum_{j=1}^{m} |b_j - a_j| \\ &< c^2 \frac{\epsilon}{c^2} = \epsilon. \end{aligned}$$

Thus f is Itô integrable to $c(B_b - B_a)$ on $[a,b]$. □

Example 2.2. Let $h \in L^2(\Omega)$. Let $s \in [a,b]$ be fixed and h be $\mathcal{F}_s$-measurable. Suppose that if $t = s$,

$$f_t(\omega) = h(\omega) \text{ for all } \omega \in \Omega$$

and if $t \neq s$,

$$f_t(\omega) = 0 \text{ for all } \omega \in \Omega.$$

Then f is Itô integrable to zero on $[a,b]$.

Proof. First assume that $E(h^2) \neq 0$. Given $\epsilon > 0$, let δ be a positive function such that $\delta(s) = \epsilon/E(h^2)$ and $\delta(\xi)$ can be defined to be any positive value for all other $\xi \in [a,b]$. Let η be arbitrarily chosen. Consider any (δ,η)-fine belated partial division $D = \{((\xi_j, v_j], \xi_j)\}_{j=1}^{m}$ of $[a,b]$. If $\xi_j \neq s$ for all $j = 1, 2, \ldots, m$, then trivially the Riemann sum

$$E\left(\sum_{j} f_{\xi_j}\left(B_{v_j} - B_{u_j}\right) - 0\right)^2 = 0.$$

Suppose we assume that $s = \xi_j$, for some j. Then,

$$\begin{aligned} E\left(\sum_j f_{\xi_i}\left(B_{v_j} - B_{\xi_j}\right) - 0\right)^2 &= E\left(f_s\left(B_{v_j} - B_{\xi_j}\right)\right)^2 \\ &= E(f_s^2)E\left(B_{v_j} - B_{\xi_j}\right)^2 \\ &= E(h^2)(v_j - \xi_j) \\ &< E(h^2)\ \epsilon/E(h^2) \\ &= \epsilon. \end{aligned}$$

Hence, f is Itô integrable to zero on $[a, b]$.

From the above proof, when $E(f_s^2) = 0$, it is clear that the result is also true. □

Observe that in Examples 2.1 and 2.2, we can choose a positive constant δ. Note that full belated δ-fine division exists if δ is constant. Hence we have the following theorem.

Theorem 2.2. *Let $f : [a, b] \times \Omega \to \mathbb{R}$ be an adapted stochastic process and $f_t \in L^2(\Omega)$ for all $t \in [a, b]$. Assume that $\{E(f_t^2) : t \in [a, b]\}$ is bounded, i.e., $|E(f_t^2)| \leq M$ for all $t \in [a, b]$. Suppose that there exists an $A \in L^2(\Omega)$ such that for each $\epsilon > 0$, there exists a positive constant δ and for any δ-fine belated full division $D = \{((\xi, v], \xi)\}$ of $(a, b]$, we have $E\left((D)\sum f_\xi(B_v - B_\xi) - A\right)^2 < \epsilon$. Then f is Itô integrable on $[a, b]$.*

Proof. Let $\epsilon > 0$ be given and let δ be a positive constant given in the theorem. We shall find $\eta > 0$ in Definition 2.5. Let $D' = \{(\xi, v]\}$ be a partial partition of $[a, b]$. Then

$$E\left((D')\sum f_\xi(B_v - B_\xi)\right)^2 = (D')\sum E(f_\xi^2)(v - \xi) \leq M(D')\sum(v - \xi).$$

Thus, if we choose a partial partition D' such that $\sum(v - \xi) < \epsilon/M$, then $E\left((D')\sum f_\xi(B_v - B_\xi)\right)^2 < \epsilon$.

Now let $\epsilon > 0$ be given, let δ be a positive constant given in the theorem. Let $\eta = \epsilon/M$. Take any (δ, η)-fine belated partial division $D = \{((\xi, v], \xi)\}$ of $[a, b]$. Let $D' = \{((\xi, v], \xi)\}$ be a δ-fine belated partial division of $[a, b]$

such that $D \cup D'$ is the full belated division of $(a, b]$. Then

$$\begin{aligned} E\left((D)\sum f_\xi(B_v - B_\xi) - A\right)^2 &\leq 2E\left((D \cup D')\sum f_\xi(B_v - B_\xi) - A\right)^2 \\ &\quad + 2E\left((D')\sum f_\xi(B_v - B_\xi)\right)^2 \\ &< 2\epsilon + 2\epsilon \\ &= 4\epsilon. \end{aligned}$$

Thus f is Itô integrable on $[a, b]$. □

We remark that by Theorem 2.2, we can use a full δ-fine belated division in Examples 2.1–2.2. Next, we shall use Theorem 2.2 to handle the next example, Example 2.3.

Example 2.3. The Brownian motion B is Itô integrable on $[a, b]$ to the value

$$\frac{1}{2}\left(B_b^2 - B_a^2\right) - \frac{1}{2}(b - a)$$

in $L^2(\Omega)$.

Proof. From Chapter 1, we know that $E(B_v - B_u)^2 = E\left(B_v^2 - B_u^2\right) = |v - u|$ and $E(B_v - B_u)^4 = E(B_v^2 - B_u^2)^2 = 3|v - u|^2$ and, for any finite collection D of disjoint intervals, we have

$$E\left(\left|(D)\sum(B_v - B_u)^2 - (v - u)\right|^2\right) = 2(D)\sum|v - u|^2.$$

Given $\epsilon > 0$. Choose $\delta(t) = 2\epsilon/b-a$ for all $t \in [a, b]$.

Note that $E(B_t^2) = t$ and $\{E(B_t^2) : t \in [a, b]\}$ is bounded. Now we shall apply Theorem 2.2 to this example.

Let $D = \{((\xi, v], \xi)\}$ be a δ-fine belated full division of $(a, b]$. Then

$$\begin{aligned} &E\left(\left|(D)\sum B_\xi(B_v - B_\xi) - \frac{1}{2}(B_b^2 - B_a^2) + \frac{1}{2}(b - a)\right|^2\right) \\ &= E\left(\left|(D)\sum\left[B_\xi(B_v - B_\xi) - \frac{1}{2}(B_v^2 - B_\xi^2) + \frac{1}{2}(v - \xi)\right]\right|^2\right) \\ &= \frac{1}{4}E\left(\left|(D)\sum(B_v - B_\xi)^2 - (v - \xi)\right|^2\right) \\ &= \frac{1}{2}(D)\sum|v - \xi|^2 \\ &< \frac{1}{2}\left(\frac{2\epsilon}{b - a}\right)(D)\sum|v - \xi| \\ &= \epsilon. \end{aligned}$$

Thus for any δ-fine belated full division of $(a,b]$,

$$E\left(\left|(D)\sum B_\xi(B_v - B_\xi) - \frac{1}{2}(B_b^2 - B_a^2) + \frac{1}{2}(b-a)\right|^2\right) < \epsilon,$$

thereby completing the proof. □

Definition 2.6. Let $a = t_0 < t_1 < t_2 < \cdots < t_n = b$ be a set of points in $[a,b]$, and $\{\psi_i\}_{i=0}^n$ be a set of random variables such that each ψ_i is $\mathcal{F}_i$-measurable and each $\psi_i \in L^2(\Omega)$.

(i) The stochastic process $f : [a,b] \times \Omega \to \mathbb{R}$ is said to be a *right-continuous* step process if it can be expressed in the form

$$f(t,\omega) = \sum_{k=1}^{n} \psi_{k-1}(\omega) 1_{[t_{k-1},t_k)}(t).$$

(ii) The stochastic process $f : [a,b] \times \Omega \to \mathbb{R}$ is said to be a *left-continuous* step process if it can be expressed in the form

$$f(t,\omega) = g(\omega)1_{\{a\}}(t) + \sum_{k=1}^{n} \psi_{k-1}(\omega) 1_{(t_{k-1},t_k]}(t),$$

where $g \in L^2(\Omega)$.

Example 2.4. Let f be either a right- or left-continuous step process in the preceding example. Then

$$\int_a^b f_t dB_t = \sum_{k=1}^{n} \psi_{k-1}\left(B_{t_k} - B_{t_{k-1}}\right).$$

It is instructional for the reader to go through the proof of left continuous step process using any positive constant δ, for each $\epsilon > 0$ and any δ-fine belated full division of $(a,b]$. Then apply the result of Example 2.2 to conclude the result of the right continuous step process. This is left as an exercise for the reader.

Example 2.5. Let $f : [a,b] \times \Omega \to \mathbb{R}$ be a process adapted to the filtering space $(\Omega, \mathcal{F}, \{\mathcal{F}_t\}, P)$ such that $E[f_t^2] = 0$ for all $t \in [a,b]$ except on a set of Lebesgue measure zero. Then f is Itô integrable on $[a,b]$ and

$$\int_a^b f_t \, dB_t = 0.$$

Proof. Let $G_k = \{t \in [a,b] : k-1 < E(f_t)^2 \leq k\}$, for $k = 1, 2, \ldots$. Then each G_k is a set of Lebesgue measure zero. Let $\epsilon > 0$ be given. For each k, there exists a countable collection $\{I_{k_j}\}$ of open intervals such that $G_k \subset \bigcup_j I_{k_j}$ and $\sum_j |I_{k_j}| < \epsilon/k2^k$. Suppose $t \in G_k$, we choose a positive number $\delta(t)$ such that whenever $(t,v] \subset (t, t+\delta(t))$, we have $(t,v] \subset \bigcup_j I_{k_j}$. If $t \notin G = \bigcup_k G_k$, the value of $\delta(t)$ can be arbitrary. Let $\eta > 0$ be any fixed positive value.

Let $D = \{((\xi, v], \xi)\}$ be a (δ, η)-fine belated partial division of $[a,b]$ and for each $k = 1, 2, \ldots$, let D_k be a subset of D such that each tag in D_k belongs to G_k. Note that some D_k may be void. Consequently,

$$E\left(\left|(D)\sum f_\xi(B_v - B_u)\right|^2\right) = \sum_{k=1}^{\infty}(D_k)\sum E(f_\xi)^2(v-\xi) < \sum_{k=1}^{\infty} k\,\frac{\epsilon}{k2^k} = \epsilon,$$

thereby completing the proof. □

Note that we cannot choose a positive constant in the above proof.

Theorem 2.3. (Uniqueness Theorem). *If f is Itô integrable on $[a,b]$, then the Itô integral of f is unique up to a set of P-measure zero, that is, if $A_1, A_2 \in L^2(\Omega)$ are both the Itô integral of f on $[a,b]$, then $A_1(\omega) = A_2(\omega)$ for all $\omega \in \Omega$ except possibly on a set $K \subset \Omega$. for which $P(K) = 0$.*

Proof. Suppose both A_1 and $A_2 \in L^2(\Omega)$ are the Itô integrals of f. Then, given $\epsilon > 0$, there exist positive functions δ_i and positive constants η_i, $i = 1, 2$, such that for any (δ_i, η_i)-fine belated partial division of $[a,b]$ denoted by $D_i = \{((\xi, v], \xi)\}$, where $i = 1, 2$, we have

$$E\left(\left|(D_i)\sum f_\xi(B_v - B_\xi) - A_i\right|^2\right) < \frac{\epsilon}{4},$$

for $i = 1, 2$. Let $\delta = \min(\delta_1, \delta_2)$ and $\eta = \min(\eta_1, \eta_2)$. Take any (δ, η)-fine belated partial division of $[a,b]$, denoting it by $D = \{((\xi, v], \xi)\}$. By definition, since $\delta \leq \delta_i$ and $\eta \leq \eta_i$ for $i = 1, 2$, this division is also (δ_i, η_i)-fine for each $i = 1, 2$. Then

$$\begin{aligned}
E\left(|A_1 - A_2|^2\right) &\leq 2E\left(\left|A_1 - (D)\sum f_\xi(B_v - B_\xi)\right|^2\right) \\
&\quad + 2E\left(\left|(D)\sum f_\xi(B_v - B_\xi) - A_2\right|^2\right) \\
&< 2\left(\frac{\epsilon}{4}\right) + 2\left(\frac{\epsilon}{4}\right) \\
&= \epsilon,
\end{aligned}$$

which means that $E\left(|A_1 - A_2|^2\right) = 0$, since $\epsilon > 0$ is arbitrary. Hence, we must have $A_1 = A_2$, possibly except on a set of P-measure zero. The proof is thereby complete. □

Notation 2.1. Subsequently, in view of the above uniqueness theorem, we shall denote the integral of the stochastic process f with respect to the Brownian motion B by the notation

$$\int_a^b f_t dB_t.$$

Thus, the notation above is well-defined except possibly for a set of P-measure zero.

In view of this notation, from Example 2.3, we can state the result as: The Brownian motion B is Itô integrable on $[a, b]$ and

$$\int_a^b B_t dB_t = \frac{1}{2}(B_b^2 - B_a^2) - \frac{1}{2}(b - a).$$

2.2 Integrable Processes

In this section, we shall establish the Cauchy Criterion. We shall also give a necessary and sufficient condition that an adapted process is Itô integrable.

Notation 2.2. For succinctness of presentation, we shall use the notation $S(f, D, \delta, \eta)$ to denote the Riemann sum

$$(D)\sum_{i=1}^{n} f_{\xi_i}\left(B_{v_i} - B_{\xi_i}\right)$$

where $D = \{((\xi_i, v_i], \xi_i)\}_{i=1}^n$ is a (δ, η)-fine belated partial division of $[a, b]$.

Theorem 2.4 (Cauchy Criterion). *Let f be an adapted process on $[a, b]$. Then f is Itô integrable on $[a, b]$ if and only if for each $\epsilon > 0$, there exist a positive function δ on $[a, b]$ and a positive constant η such that*

$$E\left(|S(f, D_1, \delta, \eta) - S(f, D_2, \delta, \eta)|^2\right) < \epsilon, \tag{2.1}$$

whenever $D_1 = \{((\xi_i, v_i], \xi_i)\}_{i=1}^n$ and $D_2 = \{((\eta_j, t_j], \eta_j)\}_{j=1}^m$ are two (δ, η)-fine belated partial divisions in $[a, b]$.

Proof. Necessity part of Cauchy Criterion follows from the inequality

$$(a - b)^2 \leq 2a^2 + 2b^2.$$

We leave it to the reader as an exercise to complete the proof.

We shall prove the sufficiency. For each $\epsilon = 1/k$, $k = 1, 2, \ldots$, let δ_k be the corresponding positive function on $[a, b]$ and η_k be the corresponding constant such that Equation (2.1) holds. Assume also that $\{\eta_k\}$ is a decreasing sequence of positive numbers and $\{\delta_k\}$ a decreasing sequence of positive functions on $[a, b]$.

For positive integers p, q with $q > p$, let D_q and D_p be the corresponding (δ_q, η_q) and (δ_p, η_p)-fine belated partial divisions in $[a, b]$, respectively. Then D_q is also (δ_p, η_p)-fine belated partial division in $[a, b]$. By (2.1),

$$E\left(|S(f, D_p, \delta_p, \eta_p) - S(f, D_q, \delta_q, \eta_q)|^2\right) < \frac{1}{p}.$$

Hence the sequence $\{S(f, D_k, \delta_k, \eta_k)\}$ is Cauchy in $L^2(\Omega)$. By the completeness of $L^2(\Omega)$, its limit under L^2-norm exists. Let $A \in L^2(\Omega)$ denote this limit. Then we have

$$E\left(|S(f, D_k, \delta_k, \eta_k) - A|^2\right) < \frac{1}{k}.$$

Given $\epsilon > 0$, choose N such that $\frac{1}{N} < \frac{\epsilon}{4}$. Let D be a (δ_N, η_N)-fine belated partial division in $[a, b]$. Then

$$\begin{aligned} E\left(|S(f, D, \delta_N, \eta_N) - A|^2\right) &\leq 2E\left(|S(f, D, \delta_N, \eta_N) - S(f, D_N, \delta_N, \eta_N)|^2\right) \\ &\quad + 2E\left(|S(f, D_N, \delta_N, \eta_N) - A|^2\right) \\ &< \frac{2}{N} + \frac{2}{N} \\ &< \epsilon, \end{aligned}$$

thereby showing that f is Itô integrable to A from Definition 2.5. Hence the proof is completed. □

Definition 2.7. A real-valued function defined on $[a, b]$ is said to be McShane integrable on $[a, b]$ if there is a number A such that for every $\epsilon > 0$, there exists a positive function δ such that for every division $D = \{([u, v], \xi)\}$ of $[a, b]$ satisfying $[u, v] \subseteq (\xi - \delta(\xi), \xi + \delta(\xi))$, we have

$$\left|(D)\sum f(\xi)(v - u) - A\right| < \epsilon.$$

Note that in the above definition we do not require $\xi \in [u,v]$. In other words, ξ may lie outside $[u,v]$. For such a division, we call it a δ-fine McShane division.

Theorem 2.5. *A real-valued function f defined on $[a,b]$ is McShane integrable on $[a,b]$ if and only if f is Lebesgue integrable on $[a,b]$.*

Definition 2.8. A real-valued function defined on $[a,b]$ is said to be McShane belated integrable on $[a,b]$ if there is a number A such that for every $\epsilon > 0$, there exist a positive function δ and a positive constant η such that for any (δ,η)-fine belated partial division $D = \{((\xi,v],\xi)\}$ of $[a,b]$, we have

$$\left|(D)\sum f(\xi)(v-\xi) - A\right| < \epsilon.$$

Theorem 2.6. *A real-valued function f defined on $[a,b]$ is McShane belated integrable on $[a,b]$ if and only if f is McShane integrable on $[a,b]$.*

Theorem 2.7. *Let f be an adapted process on $[a,b]$. Suppose that $E(f_t^2)$ exists for each $t \in [a,b]$. Then f is Itô integrable on $[a,b]$ if and only if $E(f_t^2)$ is Lebesgue integrable on $[a,b]$.*

Proof. Let $\epsilon > 0$ be given. Choose a positive function δ and a positive constant η that correspond to the Cauchy Criteria in Theorem 2.4, and let $D = \{((\xi_i,w_i],\xi_i)\}_{i=1}^{p}$ and $D' = \{((\xi_i',w_i'],\xi_i')\}_{i=1}^{q}$ be two (δ,η)-fine belated partial division of $[a,b]$. Let $\{(u_i,v_i]\}_{i=1}^{n}$ be a refinement of the two partitions from D and D', and the refined divisions are denoted by $D_1 = \{((u_i,v_i],\gamma_i)\}_{i=1}^{n}$ and $D_2 = \{((u_i,v_i],\beta_i)\}_{i=1}^{n}$, where the tags γ_i and β_i are to the left of each interval $(u_i,v_i]$.

Note that for any $u_i < v_i \le u_j < v_j$, we have

$$E\left((f(\gamma_i) - f(\beta_i))(f(\gamma_j) - f(\beta_j))(B_{v_i} - B_{u_i})(B_{v_j} - B_{u_j})\right) = 0,$$

and for $\gamma_i > \beta_i$, we have

$$E\left(f(\gamma_i) - f(\beta_i)\right)^2 = E\left(f^2(\gamma_i) - f^2(\beta_i)\right).$$

Hence

$$\begin{aligned}
E(|S(f, D,\delta, \eta) &- S(f, D', \delta, \eta)|)^2 \\
&= E(|S(f, D_1, \delta, \eta) - S(f, D_2, \delta, \eta)|)^2 \\
&= \sum_{i=1}^{n} E\left((f(\gamma_i) - f(\beta_i))^2 (B_{v_i} - B_{u_i})^2 \right) \\
&= \sum_{i=1}^{n} E\left((f^2(\gamma_i) - f^2(\beta_i)) (B_{v_i} - B_{u_i})^2 \right) \\
&= \sum_{i=1}^{n} E\left(f^2(\gamma_i) - f^2(\beta_i) \right) (v_i - u_i) \\
&= \sum_{i=1}^{n} \left(E\left(f^2(\gamma_i) \right) - E\left(f^2(\beta_i) \right) \right) (v_i - u_i) \\
&= (D_1) \sum_{i=1}^{n} E\left(f^2(\gamma_i) \right) (v_i - u_i) - (D_2) \sum_{i=1}^{n} E\left(f^2(\beta_i) \right) (v_i - u_i) \\
&= (D) \sum_{i=1}^{p} E\left(f^2(\xi_i) \right) (w_i - \xi_i) - (D') \sum_{i=1}^{q} E\left(f^2(\xi_i') \right) (w_i' - \xi_i').
\end{aligned}$$

Therefore, by Theorems 2.5 and 2.6, f is Itô integrable on $[a, b]$ if and only if $E(f_t^2)$ is Lebesgue integrable on $[a, b]$. □

2.3 Basic Properties

The following properties we are going to establish are the standard properties of integration theory.

Theorem 2.8. *Let f and g be Itô integrable on $[a, b]$, and let $\alpha \in \mathbb{R}$. Then $f \pm g$, αf are Itô integrable on $[a, b]$ and that*

(i) $\int_a^b (f_t \pm g_t) dB_t = \int_a^b f_t dB_t \pm \int_a^b g_t dB_t$
(ii) $\int_a^b \alpha f_t dB_t = \alpha \int_a^b f_t dB_t$.

Proof. We shall prove (i) first. Given $\epsilon > 0$, there exist corresponding positive functions δ_1 and δ_2 and positive constants η_1 and η_2 such that

$$E\left(\left| S(f, D_1, \delta_1, \eta_1) - \int_a^b f_t dB_t \right|^2 \right) < \frac{\epsilon}{4}$$

and

$$E\left(\left|S(g, D_2, \delta_2, \eta_2) - \int_a^b g_t dB_t\right|^2\right) < \frac{\epsilon}{4}$$

for any (δ_1, η_1)-fine belated partial division D_1 in $[a, b]$ and (δ_2, η_2)-fine belated partial division D_2 in $[a, b]$. Take $\delta = \min(\delta_1, \delta_2)$ and $\eta = \min(\eta_1, \eta_2)$. Let D be a (δ, η)-fine belated partial division in $[a, b]$, which is both (δ_1, η_1)-fine and (δ_2, η_2)-fine. Then

$$\begin{aligned}
&E\left(\left|S(f \pm g, D, \delta, \eta) - \left(\int_a^b f_t dB_t \pm \int_a^b g_t dB_t\right)\right|^2\right) \\
&\leq 2E\left(\left|S(f, D, \delta, \eta) - \int_a^b f_t dB_t\right|^2\right) + 2E\left(\left|S(g, D, \delta, \eta) - \int_a^b g_t dB_t\right|^2\right) \\
&< 2\left(\frac{\epsilon}{4}\right) + 2\left(\frac{\epsilon}{4}\right) \\
&= \epsilon,
\end{aligned}$$

showing that $f \pm g$ is Itô integrable to $\int_a^b f_t dB_t \pm \int_a^b g_t dB_t$ on $[a, b]$, thereby completing our proof.

To prove (ii): The case when $\alpha = 0$ is obvious. We shall assume that $\alpha \neq 0$. Also,

$$S(\alpha f, D, \delta, \eta) = \alpha S(f, D, \delta, \eta).$$

Hence given $\epsilon > 0$, there exist a positive function δ_1 on $[a, b]$ and a positive number η_1 such that for any (δ_1, η_1)-fine belated partial division D_1 on $[a, b]$ we have

$$E\left(\left|S(f, D_1, \delta_1, \eta_1) - \int_a^b f_t dB_t\right|^2\right) < \frac{\epsilon}{|\alpha|^2}.$$

Together with the above result, we obtain that

$$E\left(\left|S(\alpha f, D_1, \delta_1, \eta_1) - \alpha \int_a^b f_t dB_t\right|^2\right) < \epsilon,$$

thereby completing (ii) of our proof. □

Theorem 2.9. *Let f be Itô integrable on $[a, c]$ and $[c, b]$. Then f is Itô integrable on $[a, b]$ and further*

$$\int_a^b f_t dB_t = \int_a^c f_t dB_t + \int_c^b f_t dB_t.$$

Proof. Let $\epsilon > 0$ be given. There exist a positive function δ_1 on $[a, c]$ and a positive constant $\eta_1 > 0$ such that whenever D_1 is a (δ_1, η_1)-fine belated partial division in $[a, c]$, we have

$$E\left(\left|S(f, D_1, \delta_1, \eta_1) - \int_a^c f_t dB_t\right|^2\right) < \frac{\epsilon}{4}.$$

Similarly, there exist a positive function δ_2 on $[c, b]$ and a positive constant $\eta_2 > 0$ such that whenever D_2 is a (δ_2, η_2)-fine belated partial division in $[c, b]$ we have

$$E\left(\left|S(f, D_2, \delta_2, \eta_2) - \int_c^b f_t dB_t\right|^2\right) < \frac{\epsilon}{4}.$$

Define a positive function δ on $[a, b]$ as follows: $\delta(\xi) = \min(\delta_1(\xi), c - \xi)$ whenever $\xi \in [a, c)$; $\delta(\xi) = \delta_2(\xi)$ whenever $\xi \in [c, b]$, and let $\eta = \min(\eta_1, \eta_2)$.

Let D be any (δ, η)-fine belated partial division of $[a, b]$. From the construction of the function δ, it is clear that if $\xi \in [a, c)$ and $(J, \xi) \in D$, then $J \subset [a, c)$; if $\xi \in [c, b]$ and $(J, \xi) \in D$, then $J \subset [c, b]$. Let $D = D_1 \cup D_2$, where D_1 is a (δ_1, η_1)-fine belated partial division in $[a, c]$ and D_2 is a (δ_2, η_2)-fine belated partial division in $[c, b]$. Hence

$$\begin{aligned} &E\left(\left|S(f, D, \delta, \eta) - \left(\int_a^c f_t dB_t + \int_c^b g_t dB_t\right)\right|^2\right) \\ &\leq 2E\left(\left|S(f, D_1, \delta_1, \eta_1) - \int_a^c f_t dB_t\right|^2\right) \\ &\quad + 2E\left(\left|S(g, D_2, \delta_2, \eta_2) - \int_c^b g_t dB_t\right|^2\right) \\ &< 2\left(\frac{\epsilon}{4}\right) + 2\left(\frac{\epsilon}{4}\right) \\ &= \epsilon, \end{aligned}$$

and the result follows. □

Theorem 2.10. *If f is Itô integrable on $[a, b]$, then f is Itô integrable on any subinterval $[c, d]$ of $[a, b]$.*

Proof. We shall use Cauchy Criterion of Theorem 2.4 to prove this theorem.

Given $\epsilon > 0$, there exist a positive function δ on $[a,b]$ and a positive constant $\eta > 0$ such that whenever D_1 and D_2 are two (δ,η)-fine belated partial divisions of $[a,b]$, we have

$$E\left(|S(f, D_1\delta, \eta) - S(f, D_2, \delta, \eta)|^2\right) < \epsilon.$$

Let δ_1, δ_2 and δ_3 be the restrictions of δ onto $[a,c]$, $[c,d]$ and $[d,b]$, respectively. Let η_1, η_2 and η_3 be three positive constants such that $\eta_1 + \eta_2 + \eta_3 \leq \eta$. Fix P_1 and P_3 be (δ_1,η_1)-fine and (δ_3,η_3)-fine belated partial division of $[a,c]$ and $[d,b]$, respectively. Let $\mathcal{D}_1$ and $\mathcal{D}_2$ be any two (δ_2,η_2)-fine belated partial division of $[c,d]$.

Then $D_1 = P_1 \cup \mathcal{D}_1 \cup P_3$ and $D_2 = P_1 \cup \mathcal{D}_2 \cup P_3$ are two (δ,η)-fine belated partial divisions of $[a,b]$. Hence

$$\begin{aligned} &E\left(|S(f, \mathcal{D}_1, \delta_2, \eta_2) - S(f, \mathcal{D}_2, \delta_2, \eta_2)|^2\right) \\ &= E\left(|S(f, D_1, \delta, \eta) - S(f, D_2, \delta, \eta)|^2\right) < \epsilon. \end{aligned}$$

Thus f is Itô integrable on $[c,d]$ by the Cauchy Criterion of Theorem 2.4.

□

2.4 Itô-isometry

In this section we shall prove Itô-isometry.

Theorem 2.11 (Sequence of Riemann Sums). *Let f be a stochastic process on $[a,b]$. Then f is Itô integrable on $[a,b]$ if and only if there exist $A \in L^2(\Omega)$, a decreasing sequence of $\{\delta_n(\xi)\}$ of positive functions defined on $[a,b]$, and a decreasing sequence of positive numbers $\{\eta_n\}$, that is, $0 < \delta_{n+1}(\xi) < \delta_n(\xi)$ and $\eta_{n+1} < \eta_n$ for all n and all $\xi \in [a,b]$, such that we have*

$$\lim_{n\to\infty} E\left(|S(f, D_n, \delta_n, \eta_n) - A|^2\right) = 0.$$

Proof. Suppose f is Itô integrable on $[a,b]$ to $A \in L^2(\Omega)$. For each $\epsilon > 0$, there exist a positive function δ on $[a,b]$ and a positive number η such that whenever D is a (δ,η)-fine belated partial division in $[a,b]$, we have

$$E\left(|S(f, D, \delta, \eta) - A|^2\right) < \epsilon.$$

Take $\epsilon = 1/n$ for $n = 1, 2, \ldots$. Let δ_n and η_n be the corresponding positive function and positive number. We may assume $\delta_{n+1}(\xi) < \delta_n(\xi)$

for all n and $\xi \in [a,b]$, and $\eta_{n+1} < \eta_n$ for all n. Let D_n be any (δ_n, η_n)-fine belated partial division in $[a,b]$ for each n. Then we have

$$E\left(|S(f, D_n, \delta_n, \eta_n) - A|^2\right) < \frac{1}{n},$$

for each $n = 1, 2, \ldots$. Hence

$$\lim_{n\to\infty} E\left(|S(f, D_n, \delta_n, \eta_n) - A|^2\right) = 0.$$

Conversely, if there exist $A \in L^2(\Omega)$ and a decreasing sequence $\{\delta_n(\xi)\}$ of positive functions on $[a,b]$ and a decreasing sequence of positive numbers $\{\eta_n\}$ such that

$$\lim_{n\to\infty} E\left(|S(f, D_n, \delta_n, \eta_n) - A|^2\right) = 0.$$

We want to show that indeed f is Itô-integrable to A by using contradiction. Suppose that f is not Itô integrable to A on $[a,b]$. Then there exists $\epsilon > 0$ such that for every positive function δ on $[a,b]$ and every positive number η there exists a (δ, η)-fine belated partial division D in $[a,b]$ with

$$E\left|S(f, D, \delta, \eta) - A\right|^2 \geq \epsilon.$$

Hence for each δ_n and η_n given above, there exists a (δ_n, η_n)-fine belated partial division D_n in $[a,b]$ which fails to cover $[a,b]$ by at most η_n, with

$$E\left|S(f, D_n, \delta_n, \eta_n) - A\right|^2 \geq \epsilon,$$

hence it leads to a contradiction. Therefore f must be Itô integrable to A on $[a,b]$. □

Lemma 2.2. *Let f be Itô integrable on $[a,b]$. For any left-open interval $I = (u,v]$, let*

$$F(I) = \int_I f_t dB_t = \int_u^v f_t dB_t.$$

Let J and K be any two disjoint left-open subintervals of $[a,b]$. Then

(i) $E(F(J)) = 0$;
(ii) *F has the orthogonal increment property, that is,* $E(F(J)F(K)) = 0$;
(iii) $E(B(J)F(K)) = 0$;
(iv) $E(F(a,t)|\mathcal{F}_s) = F(a,s)$ *if* $s \leq t$; *and*
(v) $E\left((f_{\xi_i}(B(J) - F(J)))(f_{\xi_j}(B(K) - F(K)))\right) = 0$, *where ξ_i is the left-end point of J and ξ_j the left-end point of K.*

Proof. To prove (i), we use the orthogonal incremental property of Brownian motion, it is clear that $E[S(f, D(J), \delta, \eta)] = 0$ for any (δ, η)-fine partial division $D(J)$ of J. Taking limit, we thus have $E(F(J)) = 0$.

To prove (ii), let $(\xi_i, v_i]$ and $(\xi_j, v_j]$ be disjoint intervals from J and K, respectively. We may assume that $v_i \leq \xi_j$. Then

$$\begin{aligned}
&E\left(f_{\xi_i}(B_{v_i} - B_{\xi_i})f_{\xi_j}(B_{v_j} - B_{\xi_j})\right) \\
&= E\left(E\left(\left(f_{\xi_i}(B_{v_i} - B_{\xi_i})f_{\xi_j}(B_{v_j} - B_{\xi_j})\right)|\mathcal{F}_{\xi_j}\right)\right) \\
&= E\left(f_{\xi_i}(B_{v_i} - B_{\xi_i})f_{\xi_j}E((B_{v_j} - B_{\xi_j})|\mathcal{F}_{\xi_j})\right) \\
&= 0.
\end{aligned}$$

Hence, for any positive function δ_n and positive constant η_n, we have

$$E\left(S(f, D(I), \delta_n, \eta_n)S(f, D(J), \delta_n, \eta_n)\right) = 0,$$

where J and K are disjoint intervals of $[a, b]$ and $D(I)$ and $D(J)$ refer to the partial divisions of J and K, respectively. Furthermore, choose the sequence of positive functions $\{\delta_n\}$ and the sequence of positive numbers $\{\eta_n\}$ such that

$$\lim_{n\to\infty} E\left(|S(f, D(J), \delta_n, \eta_n) - F(J)|^2\right) = 0$$

and

$$\lim_{n\to\infty} E\left(|S(f, D(K), \delta_n, \eta_n) - F(K)|^2\right) = 0,$$

then

$$E(F(J)F(K)) = \lim_{n\to\infty} E\left(S(f, D(J), \delta_n, \eta_n)S(f, D(K), \delta_n, \eta_n)\right) = 0.$$

To prove (iii), we use the argument similar to (ii) above. We have

$$E(B(J)F(K)) = \lim_{n\to\infty} E(B(J)S(f, D(K), \delta_n, \eta_n)) = 0,$$

thereby completing proofs of (iii).

We next prove (iv). Let $s \leq t$. Define the process

$$G_t = \sum_{i=1}^{n} \eta_i(B_{v_i \wedge t} - B_{u_i \wedge t}),$$

where each η_i is $\mathcal{F}_{u_i}$-measurable, in which $u \wedge t$ is the minimum of u and t. It is clear that

$$E(G_t|\mathcal{F}_s) = G_s.$$

Thus

$$E\left(S(f,D,\delta,\eta)|\mathcal{F}_s\right) = (D)\sum f_\xi(B_{v\wedge s} - B_{u\wedge s}).$$

Hence, we have

$$E\left(F(a,b)|\mathcal{F}_s\right) = F(a,s),$$

where $a \le s \le b$. Similarly,

$$E\left(F(a,t)|\mathcal{F}_s\right) = F(a,s).$$

For conditional expectation, for example, see [Chung and Williams (1990)].

By (iii) and (iv) we get (v). □

Lemma 2.3. *Let f be Itô integrable on $[a,b]$. For any left-open interval $(u,v]$ of $[a,b]$, define $F(u,v) = \int_u^v f_t dB_t$. Let $D = \{((\xi_i, v_i], \xi_i)\}_{i=1}^n$ be a δ-fine belated partial division of $[a,b]$. Then*

(i) $E\left(\left|\sum_{i=1}^n f_{\xi_i}\left(B_{v_i} - B_{\xi_i}\right)\right|^2\right) = \sum_{i=1}^n E\left(f_{\xi_i}^2\right)(v_i - \xi_i);$

(ii) $E\left(\left|\sum_{i=1}^n f_{\xi_i}\left(B_{v_i} - B_{\xi_i}\right) - F(\xi_i, v_i)\right|^2\right)$
$= E\left(\sum_{i=1}^n \left|f_{\xi_i}\left(B_{v_i} - B_{\xi_i}\right) - F(\xi_i, v_i)\right|^2\right).$

Proof. To prove (i), note that

$$\begin{aligned}
&E\left(\left|\sum_i f_{\xi_i}(B_{v_i} - B_{\xi_i})\right|\right)^2 \\
&= E\left(\sum_i (f_{\xi_i}(B_{v_i} - B_{\xi_i}))^2 + \sum_{i\neq j}\left(f_{\xi_i} f_{\xi_j}(B_{v_i} - B_{\xi_i})(B_{v_j} - B_{\xi_j})\right)\right) \\
&= \sum_i \left(E\left(\left(f_{\xi_i}^2\right)(B_{v_i} - B_{\xi_i})^2\right)\right) \\
&= \sum_i E\left(f_{\xi_i}^2\right)(v_i - \xi_i),
\end{aligned}$$

thereby completing the proof of (i).

To prove (ii), note that

$$E\left(\left|\sum_i f_{\xi_i}(B_{v_i}-B_{\xi_i})-F(\xi_i,v_i)\right|^2\right)$$
$$=E\left(\sum_i (f_{\xi_i}B(\xi_i,v_i]-F(\xi_i,v_i))^2\right.$$
$$\left.+2\sum_{i<j}(f_{\xi_i}(B_{v_i}-B_{\xi_i})-F(\xi_i,v_i))\left(f_{\xi_j}(B_{v_j}-B_{\xi_j})-F(\xi_j,v_j)\right)\right)$$
$$=E\left(\sum_i |f_{\xi_i}(B_{v_i}-B_{\xi_i})-F(\xi_i,v_i)|^2\right)$$

which completes the proof of (ii). □

Theorem 2.12 (Itô-isometry). *Let f be Itô integrable on $[a,b]$ and $E(f_t^2)$ exists for each $t\in[a,b]$. Then $E\left(f_t^2\right)$ is Lebesgue integrable on $[a,b]$ and*

$$E\left(\int_a^b f_t dB_t\right)^2=\int_a^b E\left(f_t^2\right)dt. \tag{2.2}$$

Proof. By Theorem 2.11, there exist a decreasing sequence $\{\delta_n\}$ of positive functions on $[a,b]$ and a decreasing sequence η_n of positive numbers such that for any (δ_n,η_n)-fine belated partial division $D_n=\{((\xi_i^{(n)},v_i^{(n)}],\xi_i^{(n)})\}_{i=1}^{p(n)}$, $n=1,2,\ldots$, we have

$$\lim_{n\to\infty}E\left(\left|S(f,D_n,\delta_n,\eta_n)-\int_a^b f_t dB_t\right|^2\right)=0.$$

Hence

$$\begin{aligned}E\left(\left(\int_a^b f_t dB_t\right)^2\right)&=\lim_{n\to\infty}E\left(|S(f,D_n,\delta_n,\eta_n)|^2\right)\\&=\lim_{n\to\infty}E\left(\sum_{i=1}^{p(n)}f_{\xi_i^{(n)}}\left(B_{v_i^{(n)}}-B_{\xi_i^{(n)}}\right)\right)^2 \qquad (2.3)\\&=\lim_{n\to\infty}\sum_{i=1}^{p(n)}E\left(f^2_{\xi_i^{(n)}}\right)(v_i^{(n)}-\xi_i^{(n)}),\end{aligned}$$

noting that last step in the Equation (2.3) follows directly from part (i) of Lemma 2.3. The above equality holds for any (δ_n, η_n)-fine belated partial division D_n of $[a, b]$. Hence $E\left(f_t^2\right)$ is Lebesgue integrable on $[a, b]$ by Theorems 2.5 and 2.6 and

$$\int_a^b E\left(f_t^2\right) dt = \lim_{n\to\infty} \sum_{i=1}^{p(n)} E\left(f_{\xi_i^{(n)}}^2\right)(v_i^{(n)} - \xi_i^{(n)}).$$

By (2.3), we get (2.2), thereby completing the proof. □

Theorem 2.13. *Let f and g be Itô integrable on $[a, b]$. Then*

(i) $E\left(\int_a^b f_t dB_t\right) = 0$;

(ii) $E\left(\sum_{i=1}^{n} \int_{\xi_i}^{v_i} f_t dB_t\right)^2 = \sum_{i=1}^{n} E\left(\int_{\xi_i}^{v_i} f_t dB_t\right)^2$ *for any finite collection* $\{(\xi_i, v_i]\}_{i=1}^n$ *of disjoint subintervals of* $[a, b]$; *and*

(iii) $E\left(\left(\int_a^b f_t dB_t\right)\left(\int_a^b g_t dB_t\right)\right) = \int_a^b E\left(f_t g_t\right) dt$, *if* $E(f_t g_t)$ *exists for each* $t \in [a, b]$.

Proof. Part (i) follows directly from part (i) of Lemma 2.2. Part (ii) follows directly from part (ii) of the same lemma. We next prove part (iii). It is easy to verify that

$$E[S(f, D, \delta, \eta)S(g, D, \delta, \eta)] = E\left(\sum_{i=1}^{n} f_\xi g_\xi (B_v - B_\xi)^2\right) = \sum_{i=1}^{n} f_\xi g_\xi (v_i - \xi_i)$$

for any (δ, η)-fine belated partial division $D = \{(\xi, (\xi, v])\}$. By Theorem 2.11, there exist a sequence of positive functions $\{\delta_n\}$ and a sequence of positive constants $\{\eta_n\}$ such that for any (δ_n, η_n)-fine belated partial division of $[a, b]$, denoted by $D_n = \{((\xi^{(n)}, v^{(n)}], \xi^{(n)})\}$, we have

$$E\left(\left(\int_a^b f_t dB_t\right)\left(\int_a^b g_t dB_t\right)\right) = \lim_{n\to\infty} \sum E\left(f_{\xi^{(n)}} g_{\xi^{(n)}}\right)\left(v^{(n)} - \xi^{(n)}\right).$$

Hence, by Theorems 2.5 and 2.6, $E(fg)$ is Lebesgue integrable and the equality of (iii) holds. □

We shall next establish the martingale property of the Itô integral.

Theorem 2.14. *Let f be Itô integrable on any subinterval $[a, b]$ of $[0, \infty)$ and define $F_s = \int_0^s f_t dB_t$. Suppose $E(f_t^2)$ exists for each $t \in [a, b]$. Then the stochastic process $\{F_s : s \geq 0\}$ is a L^2-martingale with respect to the natural filtration of a Brownian motion $\{B_s : s \geq 0\}$. Furthermore, F_s is a random variable with mean 0 and variance $\int_0^s E(f_t^2)\, dt$.*

Proof. Firstly, observe that

$$E\left(F_s^2\right)=E\left(\left(\int_0^s f_t dB_t\right)^2\right)=\int_0^s E\left(f_t^2\right)dt<\infty$$

by the Itô isometry. Secondly, it is clear that F_s is $\mathcal{F}_s$-measurable for each $s\in[a,b]$. Furthermore, $\int_s^t f_u dB_u$ is independent of $\mathcal{F}_s$ if $s<t$. Finally, if $s<t$, then

$$\begin{aligned}
E(F_t|\mathcal{F}_s) &= E\left(\left(\int_0^t f_u dB_u\right)|\mathcal{F}_s\right)\\
&= E\left(\left(\int_0^s f_u dB_u\right)|\mathcal{F}_s\right)+E\left(\left(\int_s^t f_u dB_u\right)|\mathcal{F}_s\right)\\
&= \int_0^s f_u dB_u + E\left(\int_s^t f_u dB_u\right)\\
&= F_s.
\end{aligned}$$

Hence $\{F_s : s\geq 0\}$ is a L^2-martingale. F_s is a random variable with mean 0 and variance $\int_0^s E(f_t^2)\,dt$ by Theorems 2.12 and 2.13. □

2.5 Itô's Formula

Itô's Formula plays the role in stochastic calculus that the fundamental theorem of calculus plays in ordinary calculus. Not only does it relate differentiation and integration, it also provides a practical method for computation of stochastic integrals. There are several versions. In this section, we shall only present two versions of the Itô formula. First we shall state the theorem and give some examples.

Theorem 2.15 (Itô's Formula). *Let $F:\mathbb{R}\to\mathbb{R}$ be twice continuously differentiable. Suppose that*

(i) *$F'(B_t)$ is Itô integrable on $[a,b]$;*
(ii) *$E(F'(B_t))^2$ is bounded over $[a,b]$; and*
(iii) *$E(F''(B_t))^2$ is bounded over $[a,b]$.*

Then for almost all $\omega\in\Omega$, we have

$$F(B_b)-F(B_a)=\int_a^b F'(B_t)dB_t+\frac{1}{2}(R)\int_a^b F''(B_t)dt$$

where $\int_a^b F'(B_t)dB_t$ is the Itô integral of $F'(B_t)$ and $(R)\int_a^b F''(B_t)dt = (R)\int_a^b F''(B_t(\omega))dt$ is the Riemann integral.

Example 2.6. Let $F(x) = x^m$, where $m \geq 2$. Then we have $F'(x) = mx^{m-1}$ and $F''(x) = m(m-1)x^{m-2}$. Hence

(i) $F'(B_t) = mB_t^{m-1}$ and

$$E(F'(B_t))^2 = m^2 E(B_t^{2(m-1)}) = m^2 \beta t^{2(m-1)}$$

for some constant β. Thus $F'(B_t)$ is adapted and $E(F'(B_t))^2$ is continuous and bounded over $[0,s]$. Therefore, it is Riemann integrable on $[0,s]$. Thus, by Theorem 2.7, $F'(B_t)$ is Itô integrable on $[0,s]$.

(ii) Similarly, $E(F''(B_t))^2$ is continuous and bounded on $[0,s]$. Hence we have

$$B_s^m = m\int_0^s B_t^{m-1} dB_t + \frac{m(m-1)}{2}(R)\int_0^s B_t^{m-2} dt.$$

Example 2.7. Let $F(x) = e^x$. Then we have $F'(x) = F''(x) = e^x$. Hence $F'(B_t) = e^{B_t}$ is adapted and

$$E(F'(B_t))^2 = E(F''(B_t))^2 = e^{2t}.$$

Thus $F'(B_t)$ is Itô integrable on any interval $[s,t]$. By Itô's Formula, we get

$$e^{B_t} - e^{B_s} = \int_s^t e^{B_u} dB_u + \frac{1}{2}(R)\int_s^t e^{B_u} du.$$

Now we shall prove Theorem 2.15 (Itô Formula). We need to prove several lemmas first.

Lemma 2.4. *Let $g = \{g_t : t \in [a,b]\}$ be an Itô integrable process such that $E(g_t)^2$ is bounded over $[a,b]$. Then given $\epsilon > 0$, there exist a positive function δ and a positive constant η such that whenever $D = \{((\xi,v],\xi)\}$ is a δ-fine belated partial division of $[a,b]$ that fails to cover $[a,b]$ by at most η, we have*

$$E\left(\left|(D \cup D^c)\sum g_\xi (B_v - B_\xi) - \int_a^b g_t dB_t\right|^2\right) < \epsilon, \tag{2.4}$$

where $\{(\xi,v] : (\xi,v] \in D^c\}$ is the collection of all those subintervals of $[a,b]$ complement to $\{(\xi,v] : (\xi,v] \in D\}$.

Proof. Given $\epsilon > 0$, there exist a positive function δ and a positive number η such that whenever $D = \{((\xi,v],\xi)\}$ is a δ-fine belated partial division of $[a,b]$ that fails to cover $[a,b]$ by at most η, we have

$$E\left(\left|(D)\sum g_\xi (B_v - B_\xi) - \int_a^b g_t dB_t\right|^2\right) < \frac{\epsilon}{4}. \tag{2.5}$$

We may choose η small enough in the above such that whenever $D^c = \{((\xi, v], \xi)\}$ is a partial division of $[a, b]$ with $(D^c) \sum |v - \xi| < \eta$, we have

$$E\left(\left|(D^c) \sum g_\xi (B_v - B_\xi)\right|^2\right) = E\left((D^c) \sum g_\xi^2 (v - \xi)\right) < \frac{\epsilon}{4}. \tag{2.6}$$

The above can be done since $E\left(g_t^2\right)$ is bounded over $[a, b]$. Combining (2.5) and (2.6), we have our result. □

We shall denote the Riemann sum in (2.4) by $\tilde{S}(g, D \cup D^c, \delta, \eta)$ in the following lemma.

Lemma 2.5. *Let $g = \{g_t : t \in [a, b]\}$ be an Itô integrable process such that $E\left(g_t^2\right)$ is bounded over $[a, b]$. Then there exist a sequence $\{\delta_n'\}$ of positive functions and a sequence $\{\eta_n'\}$ of positive constants such that*

$$\lim_{n \to \infty} \tilde{S}(g, D_n \cup D_n^c, \delta_n', \eta_n') = \int_a^b g_t dB_t \quad \textit{in probability}, \tag{2.7}$$

where D_n is any δ_n'-fine belated partial division failing to cover $[a, b]$ by at most η_n'.

Proof. We remark that (2.7) holds for L^2-convergence by Lemma 2.4. Therefore (2.7) holds for convergence in probability. □

Lemma 2.6. *Let $h = \{h_t : t \in [a, b]\}$ be a process that for each $\omega \in \Omega$, the function $h_t(\omega)$ is continuous on $[a, b]$. Then for each $\omega \in \Omega$, there exists a sequence $\{\delta_n(\omega)\}$ of positive constants such that*

$$\lim_{n \to \infty} S^*(h, D_n, \delta_n(\omega)) = (\mathcal{R}) \int_a^b h_t(\omega) dt, \tag{2.8}$$

where $S^*(h, D_n, \delta_n(\omega)) = (D_n) \sum h_\theta (v - \xi)$ and $D_n = \{((\xi, v], \xi)\}$ is any $\delta_n(\omega)$-fine full division of $[a, b]$ and θ is any point in $[\xi, v]$.

Proof. The lemma is a consequence of the definition of the Riemann integral. We remark that in the above D_n depends on ω. □

Lemma 2.7. *Let $h = \{h_t : t \in [a, b]\}$ be an adapted process such that $E(h_t^2)$ is bounded over $[a, b]$. Then there exists a sequence $\{\delta_n\}$ of positive constants such that*

$$\lim_{n \to \infty} \hat{S}(h, D_n, \delta_n) = 0 \quad \textit{in probability}, \tag{2.9}$$

where $\hat{S}(h, D_n, \delta_n) = (D_n) \sum h_\xi ((B_v - B_\xi)^2 - (v - \xi))$ and $\{D_n\}$ is any δ_n-fine full division of $[a, b]$.

Proof. Let $|E(h_t^2)| \leq \beta$ for all $t \in [a,b]$. Let $\{\delta_n\}$ be a sequence of positive constants such that $\lim_{n\to\infty} \delta_n = 0$ and $D_n = \{((\xi,v],\xi)\}$ be any δ_n-fine full division of $[a,b]$. Since for fixed u, $(B_v - B_u)^2 - (v-u)$ is a martingale and h is adapted, we have

$$\begin{aligned}
E\Big(\Big|(D_n)\sum h_\xi\left((B_v - B_\xi)^2 - (v-\xi)\right)\Big|^2\Big) &= E\left((D_n)\sum h_\xi^2\left((B_v - B_\xi)^2 - (v-\xi)\right)^2\right) \\
&= E\left((D_n)\sum h_\xi^2 E\left(\left((B_v - B_\xi)^2 - (v-\xi)\right)^2 | \mathcal{F}_\xi\right)\right) \\
&= 2E\left((D_n)\sum h_\xi^2 (v-\xi)^2\right) \\
&< 2\delta_n E\left((D_n)\sum h_\xi^2 (v-\xi)\right) \\
&= 2\delta_n \left((D_n)\sum E(h_\xi^2)(v-\xi)\right) \\
&\leq 2\delta_n \beta (b-a).
\end{aligned}$$

So (2.9) holds for L^2-convergence, consequently it holds for convergence in probability. □

Lemma 2.8. *Let $h = \{h_t : t \in [a,b]\}$ be an adapted process such that $E(h_t^2)$ is bounded over $[a,b]$ and for each $\omega \in \Omega$, $h_t(\omega)$ is continuous on $[a,b]$. Then for each $\omega \in \Omega$, there exists a sequence $\{\delta_n(\omega)\}$ of positive constants such that for any given sequence $\{D_n(\omega)\}$ of $\delta_n(\omega)$-fine full division, we have*

$$\lim_{n\to\infty} S_w(h, D_n(\omega), \delta_n(\omega)) = (R)\int_a^b h_t(\omega)dt \quad \textit{in probability,} \tag{2.10}$$

where

$$S_\omega(h, D_n(\omega), \delta_n(\omega)) = (D_n(\omega))\sum h_\theta (B_v - B_\xi)^2$$

and θ is any point in $[\xi, v]$.

Proof. First, by Lemmas 2.6 and 2.7, (2.10) holds for $\theta = \xi$. Let $\delta_n(\omega)$ be given in Lemma 2.6 for $\theta = \xi$. We may assume that for each $\omega \in \Omega$,

$$|h_\xi(\omega) - h_\theta(\omega)| \leq \frac{1}{n}$$

whenever $\theta \in [\xi, v]$ and $((\xi,v],\xi)$ is $\delta_n(\omega)$-fine. Now let $D_n(\omega) = \{((\xi,v],\xi)\}$ be a $\delta_n(\omega)$-fine full division of $[a,b]$. Then

$$\begin{aligned}
E\Big(\Big|(D_n(\omega))\sum h_\theta (B_v - B_\xi)^2 - (D_n(\omega))\sum h_u (B_v - B_\xi)^2\Big|\Big) \\
\leq \frac{1}{n}\left[(D_n(\omega))E\left(\sum (B_v - B_\xi)^2\right)\right].
\end{aligned}$$

Consequently, (2.10) holds true. □

Proof of Theorem 2.15. Applying Taylor's Theorem to F at $B_\xi(\omega)$ for $\xi \in [a,b]$ and $\omega \in \Omega$, we have for any $v > \xi$,

$$\begin{aligned} F(B_v(\omega)) - F(B_\xi(\omega)) &= F'(B_\xi(\omega))(B_v(\omega) - B_\xi(\omega)) \\ &\quad + \frac{1}{2}F''(\alpha(\omega))(B_v(\omega) - B_\xi(\omega))^2 \end{aligned} \tag{2.11}$$

where $\alpha(\omega)$ is between $B_v(\omega)$ and $B_\xi(\omega)$. Note that for each ω, $B_t(\omega)$ is continuous on $[\xi, v]$. Hence there exists $\theta(\omega)$ between ξ and v such that $\alpha(\omega) = B_{\theta(\omega)}(\omega)$. Therefore we have

$$F(B_v) - F(B_\xi) = F'(B_\xi)(B_v - B_\xi) + \frac{1}{2}F''(B_\Theta)(B_v - B_\xi)^2. \tag{2.12}$$

For a fixed $\omega \in \Omega$, we may assume that in Lemma 2.5, each $\delta'_n \leq \delta_n(\omega)$, $\eta'_n \leq \delta_n(\omega)$, where $\delta_n(\omega)$ is given in Lemma 2.8. First, we choose $\{D_n(\omega)\}$ such that (2.7) holds in probability. In (2.7), let $u_n(\omega) = \tilde{S}(g, D_n \cup D_n^c, \delta'_n, \eta'_n)$ and $u(\omega) = \int_a^b g_t dB_t$. Then $u_n(\omega)$ converges to $u(\omega)$ in probability. Note that $D_n(\omega)$ is δ'_n-fine, so $D_n(\omega)$ is $\delta_n(\omega)$-fine. Thus (2.10) also holds in probability. In (2.10), let $v_n(\omega) = S_w(h, D_n(\omega), \delta_n(\omega))$ and $v(\omega) = (R)\int_a^b h_t(\omega)dB_t$. Then $v_n(\omega)$ converges to $v(\omega)$ in probability. Hence we can choose subsequences $\{u_{n(k)}\}$ of $\{u_n\}$ and $\{v_{n(k)}\}$ of $\{v_n\}$ such that for almost all $\omega \in \Omega$, $u_{n(k)}(\omega)$ converges to $u(\omega)$ and $v_{n(k)}(\omega)$ converges to $v(\omega)$ pointwise, respectively. Therefore we get the required result by (2.12). □

Theorem 2.16. *Let $F : \mathbb{R} \times \mathbb{R} \to \mathbb{R}$ be a function whose second order partial derivatives are continuous. Suppose that*

(i) *$F_2(t, B_t)$ is Itô integrable on $[a,b]$;*
(ii) *$E(F_2(t, B_t))^2$ is bounded over $[a,b]$; and*
(iii) *$E(F_{2,2}(t, B_t))^2$ is bounded over $[a,b]$.*

Then for almost all $\omega \in \Omega$, we have

$$F(b, B_b) - F(a, B_a) = (R)\int_a^b \left[F_1(t, B_t) + \frac{1}{2}F_{2,2}(t, B_t)\right] dt + \int_a^b F_2(t, B_t)dB_t.$$

Proof. We shall only give the ideas. Applying Taylor's Theorem to F at $(\xi, B_\xi(\omega))$ for fixed $\omega \in \Omega$, we have

$$\begin{aligned} F(v, B_v) - F(\xi, B_\xi) &\approx F_1(\xi, B_\xi)(v - \xi) + F_2(\xi, B_\xi)(B_v - B_\xi) \\ &\quad + \frac{1}{2}\big[F_{1,1}(\xi, B_\xi)(v - \xi)^2 \\ &\quad + 2F_{1,2}(\xi, B_\xi)(v - \xi)(B_v - B_\xi) \\ &\quad + F_{2,2}(\xi, B_\xi)(B_v - B_\xi)^2\big]. \end{aligned} \tag{2.13}$$

In the above B_v and B_ξ represents $B_v(\omega)$ and $B_\xi(\omega)$, respectively. Let δ be a positive function and $D = \{((\xi, v], \xi)\}$ be a δ-fine partial division. Then

$$\left|(D)\sum F_{1,1}(\xi, B_\xi(\omega)(v-\xi)^2\right| < \delta(D)\sum F_{1,1}(\xi, B_\xi(\omega))(v-\xi). \quad (2.14)$$

Note that for fixed ω, $F(t, B_t(\omega))$ is continuous on $[a, b]$. Thus $F_{1,1}(t, B_t(\omega))$ is Riemann integrable on $[a, b]$. Therefore the sum on the left-hand side of (2.14) can be made arbitrarily small. Thus Itô's Formula does not involve $F_{1,1}$. Next

$$\left|(D)\sum F_{1,2}(\xi, B_\xi)(v-\xi)(B_v(\omega) - B_\xi(\omega))\right| \leq \alpha \left|(D)\sum F_{1,2}(\xi, B_\xi)(v-\xi)\right| \quad (2.15)$$

where $|B_v(\omega) - B_\xi(\omega)| \leq \alpha$ whenever $((\xi, v], \xi)$ is δ-fine. Therefore, the sum on the left-hand side of (2.15) can also be made arbitrarily small. Thus Itô's Formula also does not involve $F_{1,2}$. Now using the same idea as in the proof of the first version of Itô Formula (Theorem 2.15) for $F_1, F_{2,2}$ and F_2, we get the required result. □

Example 2.8 (The Itô Exponential). Let $F(t, x) = e^{x-0.5t}$. A direct computation shows that $F_1(t, x) = -\frac{1}{2}F(t, x)$, $F_2(t, x) = F(t, x)$ and $F_{22}(t, x) = F(t, x)$. Apply Theorem 2.16 to $F(t, x)$, we get

$$\begin{aligned} F(b, B_b) - F(a, B_a) &= (R)\int_a^b 0dt + \int_a^b F_2(t, B_t)dB_t \\ &= \int_a^b F(t, B_t)dB_t. \end{aligned}$$

So $e^{B_b-0.5b} - e^{B_a-0.5a} = \int_a^b e^{x-0.5t}dB_t$.

Example 2.9 (Geometric Brownian Motion). Let $F(t, x) = e^{(c-0.5\sigma^2)t+\sigma x}$ where c and $\sigma > 0$ are constants. Then $F_1(t, x) = (c - 0.5\sigma^2)F(t, x)$, $F_2(t, x) = \sigma F(t, x)$ and $F_{22}(t, x) = \sigma^2 F(t, x)$. Apply Theorem 2.16 to $F(t, x)$, we get

$$F(b, B_b) - F(a, B_a) = (R)\int_a^b cF(t, B_t)dt + \int_a^b \sigma F(t, B_t)dB_t.$$

2.6 Henstock's Lemma and AC²

In this section, we shall present two versions of Henstock's Lemma, which is the standard result of Henstock's integration theory. Henstock's Lemma can be used to establish the absolute continuity property of the primitive, which shall be stated here. Subsequently, in the last theorem of this section, we shall establish a variational definition of the Itô integral by making use of Henstock's Lemma.

By Theorem 2.10, f is Itô integrable on $[a, b]$, then f is Itô integrable on any subinterval $[u, v] \subset [a, b]$. Thus, if $\int_a^b f_t dB_t$ is defined, then $F(u, v) = \int_u^v f_t dB_t$ will also be defined. We shall use this notation in the statement and proof of Henstock's Lemma below.

Lemma 2.9 (Henstock's Lemma). *Let f be Itô integrable on $[a, b]$ and $F(u, v) = \int_u^v f_t dB_t$ for any $[u, v] \subseteq [a, b]$. Then for each $\epsilon > 0$, there exists a positive function δ on $[a, b]$ such that whenever $D = \{((\xi, v], \xi)\}$ is a δ-fine belated partial division of $[a, b]$,*

$$E\left(\left|(D)\sum(f_\xi(B_v - B_\xi) - F(\xi, v))\right|^2\right) < \epsilon.$$

Note that in the statement of Henstock's Lemma above, D is *any* collection $\{((\xi_i, v_i], \xi_i)\}$ of disjoint interval-point pairs of $[a, b]$ which is δ-fine. Note that D need not cover almost the entire interval.

Proof. Let $\epsilon > 0$. Then there exist a positive function δ on $[a, b]$ and a positive number η such that whenever $\mathcal{D}$ is a (δ, η)-fine belated partial division of $[a, b]$, we have

$$E\left(\left|S(f, \mathcal{D}, \delta, \eta) - \int_a^b f_t dB_t\right|^2\right) < \epsilon.$$

Let $D = \{((\xi_i, v_i], \xi_i)\}_{i=1}^n$ be a δ-fine belated partial division of $[a, b]$. Then the set $(a, b] \setminus \bigcup_{i=1}^n (\xi_i, v_i]$ consists of a finite number of disjoint left-open intervals, say $(a_i, b_i]$, $i = 1, 2, \ldots, m$. Then by Theorem 2.10, f is Itô integrable over each $[a_i, b_i]$. Hence for each $i = 1, 2, \ldots, m$, there exist a positive function δ_i and a positive constant η_i such that whenever D_i is a (δ_i, η_i)-fine belated partial division D_i of $[a_i, b_i]$, we have

$$E\left(\left|S(f, D_i, \delta, \eta_i) - \int_{a_i}^{b_i} f_t dB_t\right|^2\right) < \frac{\epsilon}{m^2}.$$

We may assume that $\delta_i \leq \delta$ for each $i = 1, 2, \ldots, m$, and that $\sum_{i=1}^m \eta_i \leq \eta$. Then $D' = D \cup D_1 \cup \cdots \cup D_m$ is a (δ, η)-fine belated partial division of

$[a, b]$. Further

$$S(f, D', \delta, \eta) - \int_a^b f_t dB_t = (D)\sum \left(f_\xi(B_v - B_\xi) - \int_\xi^v f_t dB_t\right) + \sum_{i=1}^m \left(S(f, D_i, \delta, \eta_i) - \int_{a_i}^{b_i} f_t dB_t\right).$$

Thus

$$E\left(\left|(D)\sum\left(f_\xi(B_v - B_\xi) - \int_u^v f_t dB_t\right)\right|^2\right) < \left(\epsilon^{\frac{1}{2}} + \sum_{i=1}^m \left(\frac{\epsilon}{m^2}\right)^{\frac{1}{2}}\right)^2$$
$$= \left(\epsilon^{\frac{1}{2}} + \epsilon^{\frac{1}{2}}\right)^2.$$

Therefore

$$E\left(\left|(D)\sum\left(f_\xi(B_v - B_\xi) - \int_\xi^v f(t) dB_t\right)\right|^2\right) < 4\epsilon$$

thereby completing the proof. □

An almost immediate consequence of Henstock's Lemma is the following corollary on the continuity of the primitive process.

Corollary 2.1. *Let f be Itô integrable on $[a, b]$. Define $F_t = \int_a^t f_s dB_s$. Then F is continuous with respect to the L^2-norm.*

We leave the proof of Corollary 2.1 to the reader as an exercise.

Lemma 2.10 (Henstock's Lemma (Strong Version)). *Let f be Itô integrable on $[a, b]$. Define $F(u, v) = \int_u^v f_t dB_t$ for any interval $[u, v] \subseteq [a, b]$. Then for every $\epsilon > 0$ there exists a positive function δ on $[a, b]$ such that whenever $D = \{((\xi_i, v_i], \xi_i)\}_{i=1}^n$ is a δ-fine belated partial division of $[a, b]$ we have*

$$E\left(\sum_{i=1}^n \left|f_{\xi_i}(B_{v_i} - B_{\xi_i}) - F(\xi_i, v_i)\right|^2\right) < \epsilon.$$

Proof. This is an immediate consequence of part (ii) of Lemma 2.3 and the Henstock's Lemma in Lemma 2.9. □

In the strong version of Henstock's Lemma, the expectation of *the sum of squares* is established, compared to the earlier version of Henstock's Lemma in which *the square of the sum* of Riemann sum was established.

The strong version of Henstock's Lemma is possible by using the stochastic properties established in Lemma 2.3.

We shall proceed to establish and prove results of absolute continuity property of the primitive of the Itô integral.

Definition 2.9. Let $F = \{F_t : t \in [a,b]\}$ be a stochastic process on $[a,b]$. Then F is said to be $AC^2[a,b]$ if given $\epsilon > 0$, there exists $\eta > 0$ such that

$$E\left|\sum(F_v - F_u)\right|^2 < \epsilon$$

for any finite collection of disjoint subintervals $\{(u,v]\}$ of $[a,b]$ for which $\sum |v-u| < \eta$.

Example 2.10. The stochastic process $F_t = B_t$ is $AC^2[a,b]$.

Proof. Given $\epsilon > 0$, take $\eta = \epsilon$. Let $\{(u,v]\}$ be a finite collection of disjoint intervals of $[a,b]$ for which $\sum |v-u| < \eta$. Then

$$E\left|\sum(B_v - B_u)\right|^2 = \sum E(B_v - B_u)^2 = \sum |v-u| < \eta = \epsilon,$$

which completes the proof. □

Example 2.11. The stochastic process $F_t = B_t^2 - t$ is $AC^2[a,b]$.

We leave the proof of Example 2.11 to the reader.

The next theorem establishes the small local Riemann sum for the stochastic integrals. It will be required to show that the primitive of an Itô integrable process is $AC^2[a,b]$.

Lemma 2.11. *Let f be Itô integrable on $[a,b]$. Given $\epsilon > 0$, there exist a positive function δ on $[a,b]$ and a positive number η such that*

$$E\left(\left|(D)\sum f_\xi(B_v - B_\xi)\right|\right)^2 < \epsilon$$

for any δ-fine belated partial division $D = \{((\xi,v],\xi)\}$ with $(D)\sum|v-\xi| < \eta$.

Proof. Given $\epsilon > 0$, there exist a positive function δ on $[a,b]$ and a positive number η such that for any (δ,η)-fine belated division D of $[a,b]$, we have

$$E\left(\left|(D)\sum f_\xi(B_v - B_\xi) - \int_a^b f_t dB_t\right|\right)^2 < \frac{\epsilon}{4}.$$

Let $D_1 = \{((\xi,v],\xi)\}$ be a δ-fine belated partial division of $[a,b]$ such that $(D)\sum|v-\xi| \le \frac{\eta}{2}$. Construct a (δ,η)-fine belated partial division D_2

of $[a,b]$ which is disjoint from D_1. It is clear that $D_1 \cup D_2$ is a also (δ, η)-fine belated partial division of $[a,b]$. Hence

$$E\left(\left|(D \cup D_1)\sum f_\xi(B_v - B_\xi) - \int_a^b f_t dB_t\right|\right)^2 < \frac{\epsilon}{4}.$$

Consequently,

$$\begin{aligned} E\left(\left|(D_1)\sum f_\xi(B_v - B_\xi)\right|^2\right) &= E\left|(D_1 \cup D_2)\sum f_\xi(B_v - B_\xi) - \int_a^b f_t dB_t \right. \\ &\quad \left. + \int_a^b f_t dB_t - (D_2)\sum f_\xi(B_v - B_\xi)\right|^2 \\ &< 2\left(\frac{\epsilon}{4}\right) + 2\left(\frac{\epsilon}{4}\right) = \epsilon, \end{aligned}$$

thereby completing the proof. □

We are now ready to prove the absolute continuity of the primitive of an Itô integrable stochastic process. This is the parallel result of the classical integration theory.

Theorem 2.17. *Let f be Itô integrable on $[a,b]$. Let $\Phi(u) = \int_a^u f_t dB_t$. Then Φ is $AC^2[a,b]$.*

Proof. Let $\epsilon > 0$ be given. By Lemma 2.11, there exist a positive function δ on $[a,b]$ and a positive constant η such that whenever $D_1 = \{((\xi, v], \xi)\}$ is a δ-fine belated partial division of $[0,1]$ with $(D_1)\sum |v - \xi| \leq \eta$, we have

$$E\left(\left|(D_1)\sum f_\xi(B_v - B_\xi)\right|^2\right) < \epsilon.$$

Let $\{(a_i, b_i]\}_{i=1}^N$ be a finite collection of disjoint subintervals from $[a,b]$, where $\sum_{i=1}^N |b_i - a_i| < \eta$. By Theorem 2.10, f is Itô integrable on each $[a_i, b_i], i = 1, 2, \ldots, N$. On each subinterval $[a_i, b_i]$, choose a positive function δ_i and a positive number η_i such that

$$E\left(\left|(D_i)\sum f_\xi(B_v - B_\xi) - \int_{a_i}^{b_i} f_t dB_t\right|^2\right) < \frac{\epsilon}{2^{2i}}$$

whenever $D_i = \{((\xi, v], \xi)\}$ is a (δ_i, η_i)-fine belated partial division of $[a_i, b_i]$. Choose a sequence of positive constants $\{\eta_i\}_{i=1}^N$ such that $\sum \eta_i \leq \eta$, and further assume that $\delta_i < \delta$ for each $i = 1, 2, \ldots, N$. Now $D = \bigcup_{i=1}^N D_i$ is a δ-fine belated partial division of $[a,b]$, with

$$\sum_{i=1}^N (D_i)\sum |v - \xi| \leq \sum |b_i - a_i| < \eta,$$

so that we have

$$E\left(\left|(\sum_{i=1}^{N} D_i)\sum f_\xi(B_v - B_\xi)\right|\right)^2 < \epsilon.$$

Notice that in the L^2-normed space, if $v_i \in L^2$ for $i = 1, 2, \ldots, N$, then

$$E\left(\sum_i v_i\right)^2 = \left\|\sum_i v_i\right\|^2 \leq \left(\sum_i \|v_i\|\right)^2 = \left(\sum_i \sqrt{E(v_i^2)}\right)^2.$$

Applying this result, we have

$$\begin{aligned} &E\left(\left|\sum_i \int_{a_i}^{b_i} f_t dB_t\right|^2\right) \\ &\leq 2\left[\sum_i \sqrt{E\left(\left|\int_{a_i}^{b_i} f_t dB_t - (D_i)\sum f_\xi(B_v - B_\xi)\right|^2\right)}\right]^2 + 2\epsilon \\ &< 2\left[\sum_{i=1}^{\infty} \frac{\sqrt{\epsilon}}{2^i}\right]^2 + 2\epsilon \leq 4\epsilon. \end{aligned}$$

Thus Φ satisfies the AC^2 property on $[a, b]$. □

Now that we have established the absolute continuity of the primitive of an Itô integrable stochastic process, we shall next present an alternative definition of an Itô integrable function using the variational approach. Instead of considering the partial divisions that cover almost the entire interval $[a, b]$, we are now using *any* partial divisions of $[a, b]$.

Theorem 2.18. *Let f and F be stochastic processes on $[a, b]$. Then f is Itô integrable on $[a, b]$ to the integral F, that is,*

$$F(b) - F(a) = \int_a^b f_t dB_t$$

if and only if the process F satisfies the AC^2 property on $[a, b]$ and that for every $\epsilon > 0$, there exists a positive function δ on $[a, b]$ such that whenever $D = \{((\xi_i, v_i], \xi_i)\}_{i=1}^n$ is a δ-fine belated partial division of $[a, b]$, we have

$$E\left((D)\sum_{i=1}^{n} (f_{\xi_i}(B_{v_i} - B_{\xi_i}) - F(\xi_i, v_i))\right)^2 < \epsilon.$$

Proof. Sufficiently of the theorem follows from Henstock's Lemma and the previous result. We just need to prove the converse. Given $\epsilon > 0$, choose a positive η such that whenever $\{(u_i, v_i]\}_{i=1}^m$ is a finite collection of subintervals of $[a, b]$ with $\sum_{i=1}^m |v_i - u_i| < \eta$ we have

$$E\left(\left|\sum (F_v - F_u)\right|^2\right) < \epsilon.$$

This is possible since F satisfies the AC^2 property on $[a, b]$. Choose a (δ, η)-fine partial division $D = \{((\xi, v], \xi)\}$ of $[a, b]$. Let the part of $[a, b]$ not covered by D, which is of measure at most η, be a finite collection of subintervals $\{(s_i, t_i]\}_{i=1}^N$. Then

$$\begin{aligned}
&E\left(\left|(D)\sum f_\xi(B_v - B_\xi) - F(a, b)\right|^2\right) \\
&\leq 2E\left(\left|(D)\sum f_\xi(B_v - B_\xi) - F(\xi, v)\right|^2\right) \\
&\quad + 2E\left(\left|\sum_{i=1}^N F(s_i, t_i)\right|^2\right) \\
&< 2\epsilon + 2\epsilon = 4\epsilon,
\end{aligned}$$

thereby showing that f is Itô integrable to F on $[a, b]$. Similarly, we can show that f is Itô integrable to $F(a, v)$ on $[a, v]$ for any $v \in [a, b]$. □

We remark that we can use ideas of this section (Section 2.6) to prove Theorem 2.19:

Theorem 2.19. *(see Theorem 2.6) A real-valued function f defined on $[a, b]$ is McShane belated integrable on $[a, b]$ if and only if f is McShane integrable on $[a, b]$.*

2.7 Convergence Theorems

In this section we shall establish convergence theorems related to Itô integrals. Recall our notation in the previous section that for any process A on $[a, b]$, we can treat it as a random variable defined on all left-open intervals by letting $A(u, v]$ to denote $A_v - A_u$ or $A(u, v)$ for any subinterval $(u, v]$ of $[a, b]$.

Definition 2.10 (Variational Convergence). Let A and $A^{(n)}$, $n = 1, 2, \ldots$, be processes on the standard filtering space $(\Omega, \mathcal{F}, \{\mathcal{F}_t\}, P)$. Then $A^{(n)}$ is said to *converge variationally* to A if given $\epsilon > 0$ there exists a positive integer N such that for any finite collection of disjoint intervals $\{(u_i, v_i]\}_{i=1}^{q}$ we have

$$E\left(\sum_{i=1}^{q}\left(A^{(n)}(u_i,v_i) - A(u_i,v_i)\right)\right)^2 < \epsilon$$

for all $n \geq N$.

For variational convergence, see also Lemma 1.3, Chapter 1.

Lemma 2.12. *Let $\{f^{(n)}\}$ be a sequence of Itô integrable processes defined on $[a,b]$. Suppose that*

$$E\left(\int_a^b \left(f_t^{(n)} - f_t^{(m)}\right) dB_t\right)^2 \to 0$$

as $n, m \to \infty$. Then there exists $A_t \in L^2(\Omega)$ for every $t \in [a,b]$ with the following properties:

(i) *for every $\epsilon > 0$, there exists a positive integer N such that for every finite collection of disjoint left-open subintervals $\{(u_i, v_i]\}_{i=1}^{p}$ of $[a,b]$, we have*

$$E\left(\sum_{i=1}^{p}\left(\int_{u_i}^{v_i} f_t^{(n)} dB_t - A(u_i,v_i)\right)\right)^2$$
$$= \sum_{i=1}^{p} E\left(\int_{u_i}^{v_i} f_t^{(n)} dB_t - A(u_i,v_i)\right)^2 < \epsilon$$

whenever $n \geq N$ and where $A(u_i, v_i)$ denotes $A_{v_i} - A_{u_i}$;

(ii) *for each $\epsilon > 0$, there exists a subsequence $\{f_t^{(n_k)} : t \in [a,b]\}$ of $\{f_t^{(n)} : t \in [a,b]\}$ such that for each fixed k, for every finite collection of disjoint left-open subintervals $\{(u_i, v_i]\}_{i=1}^{p}$ of $[a,b]$, we have*

$$E\left(\left|\sum_{i=1}^{p}\left(\int_{u_i}^{v_i} f_t^{(n_k)} dB_t - A(u_i,v_i)\right)\right|^2\right)$$
$$= \sum_{i=1}^{p} E\left(\left|\int_{u_i}^{v_i} f_t^{(n_k)} dB_t - A(u_i,v_i)\right|^2\right) < \frac{\epsilon}{2^k}.$$

Proof. (i) By Theorem 2.13(ii), for each $u \in [a,b]$, we must have

$$E\left(\int_a^u \left(f_t^{(n)} - f_t^{(m)}\right) dB_t\right)^2 \to 0$$

as $n, m \to \infty$. Hence by the completeness of $L^2(\Omega)$, for each $u \in [a,b]$, there exists an $A_u \in L^2(\Omega)$ such that

$$E\left(\int_a^u f_t^{(n)} dB_t - A_u\right)^2 \to 0$$

as $n \to \infty$. Thus

$$\begin{aligned}
E&\left(\left(\int_u^v f_t^{(n)} dB_t - A(u,v)\right)^2\right) \\
&= E\left(\int_a^v f_t^{(n)} dB_t - A_v - \left(\int_a^u f_t^{(n)} dB_t - A_u\right)\right)^2 \\
&\leq 2E\left(\int_a^v f_t^{(n)} dB_t - A_v\right)^2 + 2E\left(\int_a^u f_t^{(n)} dB_t - A_u\right)^2 \to 0
\end{aligned}$$

as $n \to \infty$. On the other hand, for every $\epsilon > 0$, there exists a positive integer N such that whenever $m, n \geq N$, we have

$$E\left(\int_a^b \left(f_t^{(n)} - f_t^{(m)}\right) dB_t\right)^2 < \epsilon.$$

Observe that, by Theorem 2.13(ii), for every finite collection of disjoint left-open subintervals $\{(u_i, v_i]\}_{i=1}^p$ of $[a,b]$, we have

$$\begin{aligned}
E\left(\sum_{i=1}^p \int_{u_i}^{v_i} \left(f_t^{(n)} - f_t^{(m)}\right) dB_t\right)^2 &= \sum_{i=1}^p E\left(\int_{u_i}^{v_i} \left(f_t^{(n)} - f_t^{(m)}\right) dB_t\right)^2 \\
&\leq E\left(\int_a^b \left(f_t^{(n)} - f_t^{(m)}\right) dB_t\right)^2
\end{aligned}$$

for any n, m. Hence

$$E\left(\sum_{i=1}^p \int_{u_i}^{v_i} \left(f_t^{(n)} - f_t^{(m)}\right) dB_t\right)^2 = \sum_{i=1}^p E\left(\int_{u_i}^{v_i} \left(f_t^{(n)} - f_t^{(m)}\right) dB_t\right)^2 < \epsilon$$

for all $m, n \geq N$ and any disjoint interval $\{(u_i, v_i]\}_{i=1}^p$ of $[a,b]$. Let $m \to \infty$, by Fatou's Lemma, we get

$$\begin{aligned}
E\left(\sum_{i=1}^p \left(\int_{u_i}^{v_i} f_t^{(n)} dB_t - A(u_i, v_i)\right)\right)^2 &= \sum_{i=1}^p E\left(\int_{u_i}^{v_i} f_t^{(n)} dB_t - A(u_i, v_i)\right)^2 \\
&\leq \liminf_{m\to\infty} \sum_{i=1}^p E\left(\int_{u_i}^{v_i} \left(f_t^{(n)} - f_t^{(m)}\right) dB_t\right)^2 \\
&< \epsilon
\end{aligned}$$

for all $n \geq N$, hence completing the proof of (i).

We remark that when $A^{(n)}(u_i, v_i) = \int_{u_i}^{v_i} f_t^{(n)} dB_t$. Then $A^{(n)}$ converges variationally to A, see Definition 2.10.

(ii) By (i), for each positive integer k, we replace ϵ by $\epsilon/2^k$, then we can choose N_k such that (i) holds when $n \geq N_k$. We may assume $N_k < N_{k+1}$. Let $n_k = N_{k+1}$. Then we get a subsequence $\{f_t^{(n_k)} : t \in [a,b]\}$ of $\{f_t^{(n)} : t \in [a,b]\}$ with the required property. □

We shall next establish the Mean Convergence Theorem for Itô integral by the above lemma.

Theorem 2.20 (Mean Convergence Theorem). *Let $f^{(n)}$,$n = 1, 2, \ldots$, be a sequence of Itô integrable processes on $[a, b]$ and f be an adapted process on $[a, b]$ such that*

(i) *for almost all $t \in [a,b]$, $E\left(f_t^{(n)} - f_t\right)^2 \to 0$ as $n \to \infty$; and*

(ii) $E\left(\int_a^b \left(f_t^{(n)} - f_t^{(m)}\right) dB_t\right)^2 \to 0$ *as $n, m \to \infty$.*

Then f is Itô integrable on $[a,b]$ and

$$E\left(\int_a^b \left(f_t^{(n)} - f_t\right) dB_t\right)^2 \to 0$$

as $n \to \infty$.

Proof. The idea of the proof is standard in the theory of non-stochastic Henstock integration. First, we may assume in (i), the convergence holds for all $t \in [a, b]$. Let $\epsilon > 0$ be given. Then for any $\xi \in [a,b]$, there exists $n(\xi) > 0$ such that

$$E\left(f_\xi^{n(\xi)} - f_\xi\right)^2 < \frac{\epsilon}{3^2(b-a)}.$$

Let the stochastic process $f^{(n)}$ be Itô integrable to $A^{(n)}$ on $[a, b]$, for each $n = 1, 2, \ldots$. By Henstock's Lemma, there exists a positive function $\delta^{(n)}$ on $[a, b]$ such that for any $\delta^{(n)}$-fine belated partial divisions $D_n = \{((\xi, v], \xi)\}$ of $[a, b]$, we have

$$E\left((D_n)\sum\left(f_\xi^{(n)}[B_v - B_\xi] - A^{(n)}(\xi, v)\right)\right)^2 < \frac{\epsilon}{3^2(2^n)^2},$$

for each $n = 1, 2, \ldots$. On the other hand,

$$E\left(\int_a^b \left(f_t^{(n)} - f_t^{(m)}\right) dB_t\right)^2 \to 0$$

as $m, n \to \infty$. By the previous lemma, Lemma 2.12 and Definition 2.10, $A^{(n)}$ converges variationally to, say, A on $[a, b]$ and there exists a subsequence $\{A^{(n_k)}\}$ of $\{A^{(n)}\}$ such that

$$E\left(\sum_{i=1}^{m}\left(A^{(n_k)}(\xi_i, v_i) - A(\xi_i, v_i)\right)\right)^2 < \frac{\epsilon}{3^2(2^k)^2}.$$

In the proof that follows, we shall use the subsequence $\{A^{(n_k)}\}$ and $\{f^{(n_k)}\}$. However, for the convenience of our presentation, we denote $\{A^{(n_k)}\}$ and $\{f^{(n_k)}\}$ by $\{A^{(n)}\}$ and $\{f^{(n)}\}$, respectively. Now, let $\delta(\xi) = \delta^{n(\xi)}(\xi)$, i.e., $\delta(\xi) = \delta^{n_k(\xi)}(\xi)$, where $n(\xi)$ is given at the very beginning of the proof, and $D = \{((\xi, v], \xi)\}$ be any δ-fine belated partial division of $[a, b]$. Thus, we have

$$\begin{aligned}
&E\Big((D)\sum(f_\xi(B_v - B_\xi) - A(\xi, v))\Big)^2 \\
&= E\Big((D)\sum\left(\left(f_\xi - f_\xi^{n(\xi)}\right)(B_v - B_\xi)\right) + (D)\sum\left(A^{n(\xi)}(\xi, v) - A(\xi, v)\right) \\
&\quad + (D)\sum\left(f_\xi^{n(\xi)}(B_v - B_\xi) - A^{n(\xi)}(\xi, v)\right)\Big)^2 \\
&\le 3E\left((D)\sum\left(\left(f_\xi - f_\xi^{n(\xi)}\right)(B_v - B_\xi)\right)\right)^2 \\
&\quad + 3E\left((D)\sum\left(A^{n(\xi)}(\xi, v) - A(\xi, v)\right)\right)^2 \\
&\quad + 3E\left((D)\sum\left(f_\xi^{n(\xi)}(B_v - B_\xi) - A^{n(\xi)}(\xi, v)\right)\right)^2 \\
&= 3X + 3Y + 3Z.
\end{aligned}$$

Now,

$$\begin{aligned}
3X &= 3E\left((D)\sum\left(f_\xi - f_\xi^{n(\xi)}\right)(B_v - B_\xi)\right)^2 \\
&= 3E\left((D)\sum\left(f_\xi - f_\xi^{n(\xi)}\right)^2(v - \xi)\right)
\end{aligned}$$

since $f_\xi - f_\xi^{n(\xi)}$ is $\mathcal{F}_\xi$-measurable. Therefore,

$$\begin{aligned}
3X &= 3(D)\sum\left((v - \xi)E\left(f_\xi^{n(\xi)} - f_\xi\right)^2\right) \\
&< 3 \cdot \frac{\epsilon}{3^2(b - a)} \cdot (D)\sum_i(v - \xi) \\
&< \frac{\epsilon}{3}.
\end{aligned}$$

We have

$$
\begin{aligned}
Y^{\frac{1}{2}} &= \left(E\left((D)\sum\left(A^{n(\xi)}(\xi,v)-A(\xi,v)\right)\right)^2\right)^{\frac{1}{2}}\\
&\le \sum_j \left(E\left((D_j)\sum\left(A^{(n_j)}(\xi,v)-A(\xi,v)\right)\right)^2\right)^{\frac{1}{2}}\\
&< \sum_j \sqrt{\frac{\epsilon}{3^2(2^j)^2}}\\
&= \frac{\sqrt{\epsilon}}{3},
\end{aligned}
$$

where $D_j = \{((\xi,v],\xi) \in D : n(\xi) = n_j\}$. Recall that we denote $\{A^{n_k}\}$ by $\{A^{(n)}\}$. Therefore $n(\xi)$ represents $n_k(\xi)$. Hence $n(\xi)$ may equal to n_j, for some j and $D = \bigcup_{j=1}^{\infty} D_j$. Some D_j may be empty.

Similarly,

$$
\begin{aligned}
Z^{\frac{1}{2}} &= \left(E\left((D)\sum\left(f_\xi^{n(\xi)}(B_v-B_\xi)-A^{n(\xi)}(\xi,v)\right)\right)^2\right)^{\frac{1}{2}}\\
&\le \sum_j \left(E\left((D_j)\sum\left(f_\xi^{n_j}(B_v-B_\xi)-A^{n_j}(\xi,v)\right)\right)^2\right)^{\frac{1}{2}}\\
&< \sum_j \sqrt{\frac{\epsilon}{3^2(2^j)^2}}\\
&= \frac{\sqrt{\epsilon}}{3},
\end{aligned}
$$

thus $3X + 3Y + 3Z \le \epsilon$. It is clear that A has the AC^2 property since all $A^{(n)}$ satisfy the AC^2 property and $A^{(n)}$ converges variationally to A. Hence f is Itô stochastic integrable to A on $[a,b]$ by Theorem 2.18. □

Theorem 2.21 (Dominated Convergence Theorem). *Let $f^{(n)}$, $n = 1,2,\ldots$ be a sequence of Itô integrable processes defined on $[a,b]$. Let f and g be stochastic processes on $[a,b]$ satisfying the following properties:*

(i) *$E(f_t^{(n)} - f_t)^2 \to 0$ as $n \to \infty$ for almost all $t \in [a,b]$;*
(ii) *$|f_t^{(n)}(\omega)| \le g_t(\omega)$ for all n, for almost all $\omega \in \Omega$ and almost all $t \in [a,b]$; and that $E(g_t^2)$ is Lebesgue integrable over $[a,b]$.*

Then f is Itô integrable on $[a,b]$. Furthermore,

$$
E\left(\int_a^b \left(f_t^{(n)} - f_t\right) dB_t\right)^2 \to 0
$$

as $n \to \infty$.

Proof. By Itô isometry, $E(f_t^{(n)})^2$ is Lebesgue integrable on $[a,b]$. By (i), $E(f_t^{(n)})^2$ converges to $E(f_t)^2$ for almost all t. By Dominated Convergence Theorem for Lebesgue integral, $E(f_t)^2$ is Lebesgue integrable on $[a,b]$ and $(L)\int_a^b E(f_t^{(n)}-f_t)^2dt \to 0$ as $n\to\infty$. Hence, we have that $(L)\int_a^b E(f_t^{(n)} - f_t^{(m)})^2dt \to 0$ as $m,n\to\infty$. By the Itô-isometry again,

$$E\left(\int_a^b \left(f_t^{(n)} - f_t^{(m)}\right) dB_t\right)^2 \to 0$$

as $n,m\to\infty$. Next, apply the Mean Convergence Theorem to $\{f^{(n)}\}$, we get the required result. □

2.8 Classical Itô Integral

This section is meant for readers who have knowledge in the classical Itô integral, so that they can compare the classical theory with the current Henstock approach. Other readers may choose to skip this section. In this section we shall first outline the theory of the classical Itô integral. Our objective in this section is to show that the Henstock approach of the Itô integral is equivalent to the classical Itô integral, if integrands f are adapted and $E\left(f_t^2\right)$ are Lebesgue integrable, thereby offering an alternative approach to study the classical Itô integral.

Let $(\Omega,\mathcal{F},\{\mathcal{F}_t\},P)$ be the standard filtering space. An adapted process f on $[a,b]$ is said to be a *simple step process* on $[a,b]$ if f can be written as

$$f_t(\omega) = \alpha_0(\omega)1_{\{a\}}(t) + \sum_{i=1}^{n} \alpha_i(\omega)1_{(u_i,v_i]}(t) \tag{2.16}$$

where α_i is a $\mathcal{F}_{u_i}$-measurable bounded random variable for each $i = 1,2,\ldots,n$, α_0 is $\mathcal{F}_a$-measurable and $\{(u_i,v_i]\}_{i=1}^n$ is a finite collection of disjoint left-open intervals of $[a,b]$ with $\bigcup(u_i,v_i] = (a,b]$.

Let $\mathcal{L}_2$ be the space of functions $f:[a,b]\times\Omega\to\mathbb{R}$ such that

(i) f is $\mathcal{B}\times\mathcal{F}$-measurable, where $\mathcal{B}$ denotes the Borel σ-algebra on $[a,b]$;
(ii) $f(t,\omega)$ is $\mathcal{F}_t$-measurable; and
(iii) $\|f\|_{\mathcal{L}_2} = E\left((L)\int_a^b f_t^2dt\right) = (L)\int_a^b E\left(f_t^2\right)dt < \infty$.

It can be easily verified that $\|\cdot\|_{\mathcal{L}_2}$ is a norm. Now we shall define the classical Itô integral for $f\in\mathcal{L}_2$.

The classical Itô integral of a simple step process f given by (2.16) is defined as

$$(CI)\int_a^b f_t dB_t = \sum_{i=1}^n \alpha_i (B_{v_i} - B_{u_i}).$$

For any $f \in \mathcal{L}_2$, there exists a sequence of simple step processes $\{f^{(n)}\}$ in $\mathcal{L}_2$ which converges to f in the norm, that is,

$$E\left(\int_a^b \left(f_t^{(n)} - f_t\right)^2 dt\right) \to 0$$

as $n \to \infty$. Hence $\{f^{(n)}\}$ is a Cauchy sequence in $\mathcal{L}_2$.

By the Itô isometric property for simple step functions in the setting of the classical Itô integral, we have

$$E\left((CI)\int_a^b f_t^{(n)} dB_t - (CI)\int_a^b f_t^{(m)} dB_t\right)^2 = E\left(\int_a^b \left(f_t^{(n)} - f_t^{(m)}\right)^2 dt\right),$$

showing that $\left\{(CI)\int_a^b f_t^{(n)} dB_t\right\}$ is Cauchy in $L^2(\Omega)$. By completeness of $L^2(\Omega)$, let

$$(CI)\int_a^b f_t dB_t = \lim_{n\to\infty} (CI)\int_a^b f_t^{(n)} dB_t$$

in L^2-norm. We define the classical Itô integral of f to be $(CI)\int_a^b f_t dB_t$.

Lemma 2.13. *Let f be a simple step process as given above. Then f is both Itô integrable and classical Itô integrable and that*

$$\int_a^b f_t dB_t = (CI)\int_a^b f_t dB_t.$$

Proof. Let f be a simple step process given by (2.16). By Example 2.4, f is Itô integrable and $\int_a^b f_t dB_t = \sum_{i=1}^n \alpha_i \left(B_{v_i} - B_{u_i}\right)$. By the classical Itô integral of a simple process,

$$(CI)\int_a^b f_t dB_t = \sum_{i=1}^n \alpha_i \left(B_{v_i} - B_{u_i}\right).$$

Hence f is both Itô integrable and classical Itô integrable and

$$\int_a^b f_t dB_t = (CI)\int_a^b f_t dB_t.$$

□

Lemma 2.14. *Let $f \in \mathcal{L}_2$ and $\{f^{(n)}\}$ be a sequence of stochastic processes in $\mathcal{L}_2$ that converges to f in $\mathcal{L}_2$-norm. Let $I_u = (CI)\int_a^u f_t dB_t$ and $I_u^{(n)} = (CI)\int_a^u f_t^{(n)} dB_t$, $n = 1, 2, \ldots$. Then $I^{(n)}$ converges in mean to I.*

Proof. Let $D = \{(u, v]\}$ be a finite collection of disjoint intervals and D^c be the collection of disjoint intervals which is the complement of D. By the orthogonal increment property of $I - I^{(n)}$,

$$E\left((D)\sum\left(I(u,v) - I^{(n)}(u,v)\right)\right)^2 = (D)\sum E\left(I(u,v) - I^{(n)}(u,v)\right)^2$$
$$\leq (D \cup D^c)\sum E\left(I(u,v) - I^{(n)}(u,v)\right)^2$$
$$= E\left((D \cup D^c)\sum\left(I(u,v) - I^{(n)}(u,v)\right)\right)^2$$
$$= E(I(a,b) - I^{(n)}(a,b))^2,$$

and we know that $E(I(a,b) - I^{(n)}(a,b))^2 \to 0$ as $n \to \infty$, thereby completing our proof. □

Theorem 2.22. *Let f be classical Itô integrable on $[a, b]$. Then f is Itô integrable on $[a, b]$ and*

$$\int_a^b f_t dB_t = (CI)\int_a^b f_t dB_t.$$

Proof. Let f be classical Itô integrable on $[a, b]$. Then there exists a sequence $\{f^{(n)}\}$ of simple step processes such that $f^{(n)} \to f$ in $\mathcal{L}_2$-norm, that is,

$$E\left(\int_a^b \left(f_t^{(n)} - f_t\right)^2 dt\right) = \int_a^b E\left(f_t^{(n)} - f_t\right)^2 dt \to 0$$

as $n \to \infty$. For almost all $t \in [a, b]$, there exists a subsequence $\{n_k\}$ such that

$$E\left(f_t^{(n_k)} - f_t\right)^2 \to 0$$

as $k \to \infty$. We shall simply re-index by k and assume that

$$E\left(f_t^{(k)} - f_t\right)^2 \to 0$$

as $k \to \infty$.

By Lemma 2.13,

$$\int_a^b f_t^{(k)} dB_t = (CI)\int_a^b f_t^{(k)} dB_t$$

and the sequence $\left\{(CI)\int_a^b f_t^{(k)} dB_t\right\}$ converges in mean by Lemma 2.14.

Applying Mean Convergence Theorem, we have that f is Itô integrable and that

$$\int_a^b f_t dB_t = \lim_{n\to\infty} \int_a^b f_t^{(k)} dB_t = \lim_{n\to\infty} (CI) \int_a^b f_t^{(k)} dB_t = (CI) \int_a^b f_t dB_t$$

thereby completing our proof. □

Theorem 2.23. *Let $f : [a,b] \times \Omega \to \mathbb{R}$ be an adapted stochastic process and $E\left(f_t^2\right)$ exists for all $t \in [a,b]$. Let f be Itô integrable on $[a,b]$. Then f is classical Itô integrable on $[a,b]$ and*

$$(CI) \int_a^b f_t dB_t = \int_a^b f_t dB_t.$$

Proof. By Theorem 2.7, if f is Itô integrable on $[a,b]$, then $E\left(f_t^2\right)$ is Lebesgue integrable on $[a,b]$. Hence $f \in \mathcal{L}_2$. Therefore f is classical Itô integrable. By Theorem 2.22,

$$(CI) \int_a^b f_t dB_t = \int_a^b f_t dB_t.$$

□

2.9 Notes and Remarks

McShane, see [McShane (1969, 1974, 1984)], modified δ-fine Henstock interval-point pairs $([u,v],\xi)$ by omitting $\delta \in [u,v]$ which is used in Definition 2.8.

For a proof of Theorem 2.5, see, e.g., [Lee (1989); McShane (1984)]. The ideas in Section 2.6 can be used to prove Theorem 2.6.

The Vitali covers have been used by [McShane (1974)] to define an integral which he calls the belated integral.

For a proof of Theorem 2.6, see, e.g., [McShane (1974); Ma, Lee and Chew (1992–93)].

Most of the results using non-uniform Riemann approach in this chapter can be found in [Chew, Tay and Toh (2001–02); Toh (2001); Toh and Chew (1999, 2002, 2005, 2010, 2012); Xu and Lee (1992–93); Pop-Stojanovic (1972)].

The non-uniform Riemann approach to non-adapted stochastic processes can be found in [Boonpogkrong and Chew (2004); Chew, Huang and Wang (2004); McShane (1984); Muldowney (2012); Yang and Toh (2014, 2016)].

It is known that the Stieltjes integral $\int_a^b f dg$ exists if f is of bounded p-variation on $[a,b]$ and g is of bounded q-variation on $[a,b]$, where $\frac{1}{p} + \frac{1}{q} > 1$, see [Dudley and Norvaisa (1999); Love and Young (1938); Young (1936)] and [Monteiro, Slavik and Tvrdy (2019), p. 136]. Boonpogkrong, see [Boonpogkrong (2004, 2007); Boonpogkrong and Chew (2004–05)], proved the above Stieltjes result using the non-uniform Riemann approach without assuming the condition that f and g do not have common discontinuous points.

A Fractional Brownian motion $B_t^H : \Omega \times [0,\infty) \to \mathbb{R}$ is a process of bounded p-variation on $[a,b]$, where $p > \frac{1}{H}$, i.e., for almost all $\omega \in \Omega$, $B_t^H(\omega)$ is of bounded p-variation on $[a,b]$. We can define $\int_a^b f_t(\omega) dB_t^H(\omega)$ for almost all $\omega \in \Omega$. Hence, $f_t(\omega)$ is integrable with respect to $B_t^H(\omega)$ for almost all $\omega \in \Omega$, if for each $\omega \in \Omega$, $f_t(\omega)$ of bounded q-variation on $[a,b]$ and $\frac{1}{p} + \frac{1}{q} > 1$, with $q > 0$. Suppose $\frac{1}{H} < p < \frac{1}{H-\eta}$ where $0 < \eta < H$. Then $0 < H - \eta < \frac{1}{p} < H$. Now we shall illustrate how to choose q. Let $\frac{1}{q} = 1 - (H-\eta) = 1 - H + \eta$. Then $\frac{1}{p} + \frac{1}{q} > H - \eta + 1 - (H-\eta) = 1$ and $q > 1$. We also can choose q such that $1 > \frac{1}{q} > 1 - (H-\eta)$, then $\frac{1}{p} + \frac{1}{q} > 1$ and $q > 1$.

Now, consider $H = \frac{1}{2}$, i.e., $B_t^H = B_t$ is a Brownian motion. Then choose p such that $\frac{1}{H} = 2 < p < \frac{1}{H-\eta}$, where η is a small positive number and choose q such that $q = \frac{1}{1-H+\eta} = \frac{1}{\frac{1}{2}+\eta} = \frac{2}{1+2\eta}$, i.e., q is slightly less than 2. Hence for a Brownian motion B_t, $\int_a^b f_t(\omega) dB_t(\omega)$ exists if $f(\omega) \in BV_q[a,b]$ for almost all ω, where q is slightly less than 2.

Recall that the Itô-integral of f is defined using L^2-norm. It is not defined with respect to a Brownian path $B_t(\omega)$, where ω is fixed. It is known that if $E(f_t^2)$ is Lebesgue integrable on $[a,b]$, then f is Itô-integrable on $[a,b]$. On the other hand if we fix ω, and consider $\int_a^b f_t(\omega) dB_t(\omega)$, then we need a stronger condition on $f_t(\omega)$, the condition is that $f_t(\omega)$ is of bounded q-variation on $[a,b]$, where q is slightly less than 2.

We remark that when we consider fractional Brownian motion B_t^H, when $H \neq \frac{1}{2}$ the Stieltjes integral may be useful since we do not have the corresponding Itô-integral for B_t^H, when $H \neq \frac{1}{2}$. Recall that for the Itô-integral, the orthogonal increment property $E\left((B_v - B_u)(B_t - B_s)\right) = 0$, where $u < v \leq s < t$, plays an important role. However, this property does not hold for B_t^H, when $H \neq \frac{1}{2}$. Therefore, we do not have the corresponding Itô integral for B_t^H, when $H \neq \frac{1}{2}$.

The above remarks about $\{B_t^H\}$ is taken from [Boonpogkrong (2004), pp. 98–100].

Chapter 3

Differentiation and Differential

This chapter consists of two parts. In the first part, we will define the belated derivative of a stochastic process and thereby characterize the class of all Itô integrable processes on $[a, b]$ by its primitive process.

The notion of differential has been used extensively in stochastic analysis and mathematical finance. However, a rigorous definition of the differential is not given.

In the second part, we provide a definition of the differential of a stochastic process and a rigorous computation for the differential of the function of a stochastic process.

3.1 Differentiation

In this section, we shall define the derivatives $D_\beta F_t$ of a process $F = \{F_t : t \in [a, b]\}$ with respect to $B = \{B_t : t \in [a, b]\}$. We shall concern ourselves with questions like: when does

$$\int_a^b D_\beta F_t dB_t = F_b - F_a?$$

If $F_x = \int_a^x f_t dB_t$, do we have $D_\beta F_t = f_t$? Here we shall also make use of belated δ-fine interval-point pair in the definition of our derivative.

Definition 3.1. Let $F = \{F_t : t \in [a, b]\}$ be an L^2-martingale. Then the process F is said to be *differentiable* at $t \in [a, b)$ if there exists a random variable f_t such that for any $\epsilon > 0$ there exists a positive number $\delta > 0$ such that whenever $(t, v] \subset [t, t + \delta(t)]$, we have

$$E\left(|f_t(B_v - B_t) - (F_v - F_t)|^2\right) < \epsilon E(B_v - B_t)^2 = \epsilon|v - t|.$$

The random variable f_t is said to be the *belated derivative* of F at the point t. We shall denote f_t by $D_\beta F_t$ in our subsequent presentation.

Theorem 3.1. *Let the adapted process f be Itô integrable on $[a,b]$ and let $F_t = \int_a^t f_s dB_s$. Then*

(i) *F is an L^2-martingale and has the AC^2 property; and*
(ii) *$D_\beta F_t = f_t$ a.e. on $[a,b)$.*

Proof. (i) follows directly from Theorems 2.14 and 2.17.

To prove (ii), we need to show that the set of points B of $[a,b)$ for which $D_\beta F_t$ does not exist or unequal to f is of Lebesgue measure zero. Let $t \in B$. By definition, there exists $\gamma(t) > 0$ such that for any positive number $\delta(t)$, there exists $(t,v] \subset [t, t+\delta(t)]$ and

$$E\left(|f_t(B_v - B_t) - (F_v - F_t)|^2\right) \geq \gamma(t)(v-t). \tag{3.1}$$

By Henstock's Lemma (Lemma 2.10), given $\epsilon > 0$, there exists a positive function β on $[a,b]$ such that whenever $D = \{((\xi, v], \xi)\}$ is a β-fine belated partial division of $[a,b]$, we have

$$E\left((D)\sum |f_\xi(B_v - B_\xi) - (F_v - F_\xi)|^2\right) < \epsilon. \tag{3.2}$$

Now we consider a special D such that each $(\xi, v]$ satisfies (3.1) and (3.2). Let the set $B_m = \{t \in [a,b] : \gamma(t) \geq \frac{1}{m}\}$, $m = 1, 2, \dots$ and fix B_m. Furthermore, we assume that each $\xi \in B_m$. Then by (3.1) and (3.2), we have

$$(D)\sum (v - \xi) < m\epsilon.$$

Let $\mathcal{G}$ be a family of collections of intervals $[\xi, v]$ induced from all β-fine belated partial division with $\xi \in B_m$ satisfying (3.1). Then $\mathcal{G}$ covers B_m in the Vitali's sense. Applying the Vitali Covering theorem, there exists a finite collection of such intervals $\{[\xi_i, t_i]\}_{i=1}^q$ such that

$$\mu(B_m) \leq \sum_{i=1}^q |t_i - \xi_i| + \epsilon < (m+1)\epsilon.$$

Hence $\mu(B_m) = 0$ and so $\mu(B) = 0$. Therefore our proof is completed. □

Theorem 3.2. *Let f be an adapted process on $[a,b]$ such that*

(i) *F is an L^2-martingale and has the AC^2 property;*
(ii) *$D_\beta F_t = f_t$ a.e. on $[a,b)$;*

then f is Itô integrable on $[a,b]$ with $F_t = \int_a^t f_s dB_s$.

Proof. Let $D_\beta F_t = f_t$ for all $t \in [a,b)$ except possibly on a set B which has Lebesgue measure zero. Let $\xi \in [a,b) \setminus B$. Given $\epsilon > 0$ there exists a positive function δ on $[a,b] \setminus B$ such that whenever $(\xi, v]$ is δ-fine, we have

$$E\left(|f_\xi(B_v - B_\xi) - (F_v - F_\xi)|^2\right) < \epsilon|v-u|.$$

Let $D = \{((\xi_i, v_i], \xi_i)\}_{i=1}^n$ be a δ-fine belated partial division of $[a,b]$ with all $\xi_i \in [a,b]\setminus B$. Then, by (i),

$$\begin{aligned}
&E\left(\left|\sum_{i=1}^n f_{\xi_i}(B_{v_i} - B_{\xi_i}) - (F_{v_i} - F_{\xi_i})\right|^2\right) \\
&= E\left(\sum_{i=1}^n |f_{\xi_i}(B_{v_i} - B_{\xi_i}) - (F_{v_i} - F_{\xi_i})|^2\right) \\
&< \epsilon \sum_{i=1}^n |v_i - \xi_i| \\
&\leq \epsilon(b-a).
\end{aligned}$$

Thus if $B = \phi$, it is clear from the above and Theorem 2.18 that f is Itô integrable with $F_t = \int_a^t f_t dB_t$. In general, B is nonempty with $\mu(B) = 0$. The following technique is used in Theory of Henstock integration. Now we let

$$B_m = \left\{t \in B : m-1 < E\left(f_t^2\right) \leq m\right\},$$

where $\mu(B_m) = 0$ and $m = 1, 2, \ldots$. Then $B = \bigcup_{m=1}^\infty B_m$.

Since F has AC^2 property, given any positive integer m, there exists $\eta_m > 0$ with $\eta_m < \frac{\epsilon}{2m\, 2^m}$ such that whenever $\{(u_i, v_i]\}_{i=1}^n$ is a finite collection of disjoint left-open subintervals of $[a,b]$ with $\sum_{i=1}^n |v_i - u_i| < \eta_m$, we have

$$E\left(\left|\sum_{i=1}^n (F_{v_i} - F_{u_i})\right|^2\right) < \frac{\epsilon}{2^m}.$$

Take an open set $G_m \supset B_m$ such that $\mu(G_m) < \eta_m$. Now, for each m, define a positive function δ on B_m as follows. Let $t \in B_m$ and define $\delta(t) > 0$ such that whenever $((t,v],t)$ is a belated δ-fine with $t \in B_m$ we have $(t,v] \subset G_m$.

Fix a positive integer m. Let $D = \{((\xi_i, v_i], \xi_i)\}_{i=1}^q$ be a δ-fine belated partial division of $[a,b)$ such that $\xi_i \in B_m$ for all i. Then we have

$$\begin{aligned}
&E\left(\left|\sum_{i=1}^{q}(f_{\xi_i}(B_{v_i}-B_{\xi_i})-(F_{v_i}-F_{\xi_i}))\right|^2\right)\\
&\leq 2E\left(\left|\sum_{i=1}^{q}(f_{\xi_i}(B_{v_i}-B_{u_i}))\right|^2\right)+2E\left(\sum_{i=1}^{q}(F_{v_i}-F_{\xi_i})^2\right)\\
&< 2\sum_{i=1}^{q}E\left(f_{\xi_i}^2\right)(v_i-u_i)+2\left(\frac{\epsilon}{2^m}\right)\\
&< 2m\left(\frac{\epsilon}{2m\,2^m}\right)+2\left(\frac{\epsilon}{2^m}\right)\\
&= 3\left(\frac{\epsilon}{2^m}\right).
\end{aligned}$$

So, considering any δ-fine belated partial division over $[a,b]$, denoted by $D_1 = \{((\xi_i, v_i], \xi_i)\}_{i=1}^{s}$, we have

$$\begin{aligned}
&E\left(\left|\sum_{i=1}^{s}(f_{\xi_i}(B_{v_i}-B_{\xi_i})-(F_{v_i}-F_{\xi_i}))\right|^2\right)\\
&\qquad\leq 2E\left(\left|\sum_{\xi_i\in[a,b]\setminus B}(f_{\xi_i}(B_{v_i}-B_{\xi_i})-(F_{v_i}-F_{\xi_i}))\right|^2\right)\\
&\qquad\quad+2E\left(\left|\sum_{m=1}^{\infty}\sum_{\xi_i\in B_m}(f_{\xi_i}(B_{v_i}-B_{\xi_i})-(F_{v_i}-F_{\xi_i}))\right|^2\right)\\
&\qquad< 2\epsilon(b-a)+2\sum_{m=1}^{\infty}3\left(\frac{\epsilon}{2^m}\right)\\
&\qquad= 2\epsilon(b-a)+6\epsilon,
\end{aligned}$$

thereby showing that f is Itô integrable with $F_t = \int_a^t f_t dB_t$ by Theorem 2.18. □

3.2 Differential

In this section, we shall define differential for $f(X_t)$, where X_t is a stochastic process with $E(\Delta X_t)^2 = o(\Delta t)$ and f has a continuous second derivative. In the deterministic case, we use Taylor's formula up to first degree-polynomial to define differential $df(x)$, and use differential to approximate the small change $f(x+\Delta x)-f(x)$. The degree of error is $o(\Delta x)$. However, for stochastic processes in this section, we use Taylor's formula up to second degree-polynomial to define differential $df(X_t)$, and use differential to approximate the small change $f(X_{t+\Delta t})-f(X_t)$. We use second degree-polynomial since we want the degree of error to be $o(\Delta t)$.

In this section, we fix a standard filtering space $(\Omega, \mathcal{F}, \{\mathcal{F}_t\}, P)$, and let the canonical Brownian motion denoted by $B_s(\omega)$ be adapted to the standard filtering space. The set of all real numbers is denoted by $\mathbb{R}$.

3.2.1 *Differentials for process* $\{X_t\}$ *with* $E(\Delta X_t)^2 = o(\Delta t)$

Let $X : [a,b]\times\Omega \to \mathbb{R}$ be a process and $X(t,\omega)$ denoted by $X_t(\omega)$. Let X be a process with $E(\Delta X_t)^2 = o(\Delta t)$, where $\Delta X_t = X_{t+\Delta t} - X_t$. The differential of X, denoted by dX, is defined as $dX_t = \Delta X_t$.

3.2.2 *Itô's formula in differential form*

Recall that for a deterministic real-valued function f, if $f'(x)$ exists at a point x, then the differential of f is defined as follows:

$$dy = f'(x)\Delta x = f'(x)dx$$

where $y = f(x)$ and dx denotes Δx.

On the other hand,

$$\Delta y = f(x+\Delta x) - f(x) = f'(x)(\Delta x) + o(\Delta x), \tag{3.3}$$

i.e., for any $\epsilon > 0$, there exists $\delta(x) > 0$ such that

$$|f(x+\Delta x) - f(x) - f'(x)(\Delta x)| < \epsilon\,|\Delta x|$$

whenever $|\Delta x| = |x + \Delta x - x| < \delta(x)$.

In dealing with classical integrals, the degree of error term $o(\Delta x)$ in (3.3) is sufficient to get the exact value of integrals over an interval $[a,b]$. We assume that the Riemann integral of $f'(t)$ exists over $[a,b]$. We take the Riemann sum $\sum$ over the interval $[a,b]$ using a δ-fine partition, where δ is given in (3.3), we get

$$\begin{aligned}\left|f(b) - f(a) - \sum f'(x)(\Delta x)\right| &= \left|\sum (f(x+\Delta x) - f(x)) - \sum f'(x)(\Delta x)\right| \\ &\leq \sum |f(x+\Delta x) - f(x) - f'(x)(\Delta x)| \\ &< \epsilon \sum |\Delta x| = \epsilon(b-a).\end{aligned} \tag{3.4}$$

The total jump $\sum |\Delta x|$ of Δx over the interval $[a, b]$ is the length $|b - a|$ of the interval, which is finite. Thus,

$$\left| f(b) - f(a) - \lim_{\Delta x \to 0} \sum f'(x)(\Delta x) \right| < \epsilon(b - a),$$

i.e.,

$$\left| f(b) - f(a) - \int_a^b f'(x)\, dx \right| < \epsilon(b - a).$$

Hence $f(b) - f(a) = \int_a^b f'(x)\, dx$. Therefore, this is another good reason to use the differential

$$dy = f'(x)\, dx.$$

By integrating both sides, we get

$$\int_a^b dy = \int_a^b f'(x)\, dx,$$

i.e.,

$$f(b) - f(a) = \int_a^b f'(x)\, dx.$$

Now suppose in (3.3), we replace $f(x+\Delta x) - f(x)$ by $f(B_{t+\Delta t}) - f(B_t)$. Then we have

$$\Delta y = f(B_{t+\Delta t}) - f(B_t) = f'(B_t)\Delta B_t + o(|\Delta B_t|),$$

where $\Delta B_t = B_{t+\Delta t} - B_t$.

It is well-known that the paths of a Brownian motion are of unbounded variation, i.e., $\sum |\Delta B_t|$ is unbounded. Hence we cannot have (3.4). Note that $\Delta y = f(B_{t+\Delta t}) - f(B_t) = f'(B_t)\Delta B_t + o(|\Delta B_t|)$ is a random variable. By taking expectations,

$$E\left(|\Delta y - f'(B_t)\Delta B_t|\right) = o(E(|\Delta B_t|)) = o(\sqrt{\Delta t}),$$

where $E(g) = \int_\Omega g\, dP$ for any random variable g. This also does not work, since the total jump of $\sqrt{\Delta t}$ is infinite. Note that $E(|\Delta B_t|) = \sqrt{\frac{2}{\pi}}\sqrt{\Delta t}$.

In the following, we assume that f has a continuous second derivative. By applying the Taylor's formula up to second degree-polynomial to a function f, we have $f(y) - f(x) = f'(x)(y-x) + \frac{1}{2}f''(x)(y-x)^2 + R(x, y)$, where $R(x, y) = \int_x^y (y-u)(f''(u) - f''(x))du$. We assume f'' is continuous. Therefore, for $\epsilon > 0$, there exists $\delta(x) > 0$ such that whenever $|y - x| < \delta(x)$ we have $\left|f(y) - f(x) - f'(x)(y - x) - \frac{1}{2}f''(x)(y - x)^2\right| < \epsilon|y - x|^2$, see, for example, [Steele (2001), p. 112]. Let $y = B_{t+\Delta t}$ and $x = B_t$, we have, for every $\epsilon > 0$, there exists a positive function $\delta(t,\omega)$ defined on $[a,b] \times \Omega$ such that

$$\left| f(B_{t+\Delta t}) - f(B_t) - f'(B_t)\Delta B_t - \frac{1}{2} f''(B_t)(\Delta B_t)^2 \right| < \epsilon \left| (\Delta B_t)^2 \right|,$$

whenever $|\Delta B_t| < \delta(t,\omega)$. Hence, for every $\epsilon > 0$, there exists $\delta(t,\omega) > 0$ such that

$$E\left(\left| f(B_{t+\Delta t}) - f(B_t) - f'(B_t)\Delta B_t - \frac{1}{2} f''(B_t)(\Delta B_t)^2 \right| \right) < \epsilon E\left(\left| (\Delta B_t)^2 \right| \right) = \epsilon |\Delta t|, \quad (3.5)$$

whenever $|\Delta B_t| < \delta(t,\omega)$. Note that the total jump of Δt over $[a,b]$ is $b-a$.

The approach used is Riemann sums with non-uniform meshes. Theorems 3.7 and 3.8 show that we can have full divisions of non-uniform meshes. Hence, by (3.5), we are able to define stochastic differential, see Definition 3.4.

Now we shall introduce $(dB_t)^2$ integrals.

Definition 3.2. Let $f : [a,b] \times \Omega \to \mathbb{R}$ be an adapted stochastic process. Then f is said to be $(dB_t)^2$ integrable on $[a,b]$ if there exists an $A \in L(\Omega)$ such that for any $\epsilon > 0$, there exist a positive function $\delta(t,\omega)$ on $[a,b] \times \Omega$ and a positive number $\eta(\omega) > 0$ such that for any (δ,η)-fine belated partial division $D = \{((\xi_i, v_i], \xi_i)\}_{i=1}^{n}$ of $[a,b]$, we have

$$E\left| \sum_{i=1}^{n} f_{\xi_i} \left(B_{v_i} - B_{\xi_i}\right)^2 - A \right| < \epsilon.$$

Here $D = \{((\xi, v], \xi)\}$ is (δ,η)-fine, where $v = v(\omega)$ if $(\xi, v(\omega)] \subset (\xi, \xi + \delta(\xi,\omega))$ and $|(b-a) - (D)\sum |v(\omega) - \xi|| < \eta(\omega)$. Furthermore, $B_v - B_\xi$, f_ξ, and A denote $B(v(\omega),\omega) - B(\xi,\omega)$, $f(\xi,\omega)$ and $A(\omega)$, respectively.

Theorem 3.3. *[Chapter 2, Theorem 2.7] Let $f : [a,b] \times \Omega \to \mathbb{R}$ be an adapted process. Suppose that $E(f_t^2)$ exists for each $t \in [a,b]$. Then f is Itô integrable on $[a,b]$ if and only if $E(f_t^2)$ is Lebesgue integrable on $[a,b]$.*

Theorem 3.4. *[Chapter 2, Theorem 2.5] A real-valued function f defined on $[a,b]$ is McShane integrable on $[a,b]$ if and only if f is Lebesgue integrable on $[a,b]$.*

Theorem 3.5. *[Chapter 2, Theorem 2.6] A real-valued function f defined on $[a,b]$ is McShane belated integrable on $[a,b]$ if and only if f is McShane integrable on $[a,b]$.*

Definition 3.3. Let $f : [a, b] \times \Omega \to \mathbb{R}$ be an adapted process. A process f is said to be McShane belated integrable to $A \in L(\Omega)$ on $[a, b]$ in $L(\Omega)$ if for any $\epsilon > 0$, there exist $\delta(t, \omega) > 0$ and $\eta(\omega) > 0$ such that for any (δ, η)-fine belated partial division $D = \{((\xi, v], \xi)\}$ of $[a, b]$, we have

$$E\left(\left|(D)\sum f_\xi(\omega)(v-\xi) - A\right|\right) < \epsilon.$$

Remark 3.1. Let $f : [a, b] \times \Omega \to \mathbb{R}$ be an adapted process. Suppose for each $\omega \in \Omega$, $f_t(\omega)$ is McShane belated (Lebesgue) integrable to $A(\omega)$ on $[a, b]$. Furthermore, we assume that $A(\omega) \in L(\Omega)$ and for each $t \in [a, b]$, $E(f_t)$ exists. Then f is McShane belated integrable to $A(\omega) \in L(\Omega)$ on $[a, b]$ in $L(\Omega)$.

Theorem 3.6. *Let $f : [a, b] \times \Omega \to \mathbb{R}$ be an adapted process. Then*

(i) *f is $(dB_t)^2$ integrable on $[a, b]$ if and only if f is McShane belated integrable on $[a, b]$ in $L(\Omega)$; and*
(ii) *suppose, for each $\omega \in \Omega$, $f_t(\omega)$ is Lebesgue (McShane belated) integrable to $(L)\int_a^b f_t(\omega)dt$ on $[a, b]$. We assume further that $E\left((L)\int_a^b f_t(\omega)dt\right)$ exists, and for each $t \in [a, b]$, $E(f_t)$ exists. Then f is $(dB_t)^2$ integrable on $[a, b]$ and*

$$E\left(\int_a^b f_t(\omega)(dB_t)^2\right) = E\left((L)\int_a^b f_t(\omega)dt\right) = (L)\int_a^b E(f_t)dt.$$

Proof. (i) Let $D = \{((\xi_j, w_j], \xi_j)\}_{j=1}^p$ and $D' = \{((\xi'_k, w'_k], \xi'_k)\}_{k=1}^q$ be two (δ, η)-fine belated partial division of $[a, b]$. Let $\{(u_i, v_i]\}_{i=1}^n$ be a refinement of two partitions from D and D' and the refined division are denoted by $D_1 = \{((u_i, v_i], \gamma_i)\}_{i=1}^n$ and $D_2 = \{((u_i, v_i], \beta_i)\}_{i=1}^n$, where the tags γ_i and β_i are to the left of each interval $(u_i, v_i]$.

Let

$$S(f, D, \delta, \eta) = (D)\sum_{j=1}^{p} f_{\xi_j}\left(B_{w_j} - B_{\xi_j}\right)^2$$

and

$$S(f, D', \delta, \eta) = (D')\sum_{k=1}^{q} f_{\xi'_j}\left(B_{w'_j} - B_{\xi'_j}\right)^2.$$

In the above, if $\left(B_{w_j} - B_{\xi_j}\right)^2$ and $\left(B_{w'_j} - B_{\xi'_j}\right)^2$ are replaced by $(w_j - \xi_j)$ and $\left(w'_j - \xi'_j\right)$, respectively, then $S(f, D, \delta, \eta)$ and $S(f, D', \delta, \eta)$ will be replaced by $S_1(f, D, \delta, \eta)$ and $S_1(f, D', \delta, \eta)$, respectively.

In the following, f_t is denoted by $f(t)$.

$$
\begin{aligned}
&E(|S(f,D,\delta,\eta) - S(f,D',\delta,\eta)|) \\
&\quad = E\left(|S(f,D_1,\delta,\eta) - S(f,D_2,\delta,\eta)|\right) \\
&\quad = E\left(\left|\sum_{i=1}^{n}(f(\gamma_i) - f(\beta_i))\left(B_{v_i} - B_{u_i}\right)^2\right|\right) \\
&\quad = E\left(\left|\sum_{i=1}^{n} E\left((f(\gamma_i) - f(\beta_i))\left(B_{v_i} - B_{u_i}\right)^2 \,\middle|\, \mathcal{F}_u\right)\right|\right) \\
&\quad = E\left(\left|\sum_{i=1}^{n} (f(\gamma_i) - f(\beta_i))\, E\left(\left(B_{v_i} - B_{u_i}\right)^2 \,\middle|\, \mathcal{F}_u\right)\right|\right) \\
&\quad = E\left(\left|\sum_{i=1}^{n} (f(\gamma_i) - f(\beta_i))\,(v_i - u_i)\right|\right) \\
&\quad = E\left(\left|\sum_{i=1}^{n} f(\gamma_i)(v_i - u_i) - f(\beta_i)(v_i - u_i)\right|\right) \\
&\quad = E\left(|S_1(f,D_1,\delta,\eta) - S_1(f,D_2,\delta,\eta)|\right) \\
&\quad = E\left(|S_1(f,D,\delta,\eta) - S_1(f,D',\delta,\eta)|\right).
\end{aligned}
$$

Hence, by Cauchy's Criterion, we get the required result.

(ii) By the conditions given in (b) and Remark 3.1, f is $(dB_t)^2$ integrable. We can find a sequence $\{D_n\}$ of $(\delta(t,\omega) - \eta(\omega))$-fine belated partial divisions of $[a,b]$ such that

$$
\begin{aligned}
E\left(\int_a^b f\,(dB_t)^2\right) &= \lim_{n\to\infty} E\left((D_n)\sum f_\xi (B_v - B_\xi)^2\right) \\
&= \lim_{n\to\infty} E\left((D_n)\sum f_\xi (v - \xi)\right) \\
&= \lim_{n\to\infty} (D_n)\sum E(f_\xi)(v - \xi) \\
&= (L)\int_a^b E\left(f_t\right) dt \\
&= E\left((L)\int_a^b f_t(\omega)dt\right).
\end{aligned}
$$

□

Theorem 3.7. *[Toh and Chew (2010)] and [Chapter 2, Lemma 2.5] Let $f : [a,b] \times \Omega \to \mathbb{R}$ be an adapted process. Suppose f is Itô integrable on $[a,b]$ and $E(f_t)^2$ exists for each $t \in [a,b]$. Then given $\epsilon > 0$, there exists a positive function $\delta(t)$ and a positive constant η such that whenever $D = \{((\xi, v], \xi)\}$ is a (δ, η)-fine belated partial division of $[a,b]$, we have*

$$E\left(\left|(D \cup D^c)\sum f_\xi (B_v - B_\xi) - \int_a^b f_t dB_t\right|\right) < \epsilon, \tag{3.6}$$

where $\{(\xi, v] : (\xi, v] \in D^c\}$ is the collection of all those subintervals of $[a,b]$ complement to $\{(\xi, v] : (\xi, v] \in D\}$.

We remark that in [Toh and Chew (2010)], the following is proved:

$$E\left(\left|(D \cup D^c)\sum f_\xi (B_v - B_\xi) - \int_a^b f_t dB_t\right|^2\right) < \epsilon.$$

Hence the inequality in Theorem 3.7 is true.

By Theorem 3.6, the $(dB_t)^2$ integral is absolutely continuous. Hence, using the ideas of the proof of Theorem 3.7, we have

Theorem 3.8. *Let $f : [a,b] \times \Omega \to \mathbb{R}$ be an adapted process. Suppose f is $(dB)^2$ integrable on $[a,b]$. Then given $\epsilon > 0$, there exist $\delta(t,\omega) > 0$ and $\eta(\omega) > 0$ such that whenever $D = \{((\xi, v], \xi)\}$ is a (δ, η)-fine belated partial division of $[a,b]$, we have*

$$E\left(\left|(D \cup D^c)\sum f_\xi (B_v - B_\xi)^2 - \int_a^b f_t (dB)^2\right|\right) < \epsilon, \tag{3.7}$$

where $\{(\xi, v] : (\xi, v] \in D^c\}$ is the collection of all those subintervals of $[a,b]$ complement to $\{(\xi, v] : (\xi, v] \in D\}$.

By Theorems 3.7 and 3.8, for any $\epsilon > 0$, there exist $\delta(t,\omega) > 0$ and $\eta(\omega) > 0$ such that (3.6) and (3.7) hold. Note that we may assume $\delta(t,\omega) \le \delta(t)$, and $\eta(\omega) \le \eta$ for all $\omega \in \Omega$, where $\delta(t,\omega)$, $\delta(\omega)$ are given in Theorem 3.8 and $\delta(t)$, η are given in Theorem 3.7. Hence, we have

$$E\left(\left|f(B_b) - f(B_a) - (D \cup D^c)\sum\left(f'(B_t)\Delta B_t + \frac{1}{2}f''(B_t)(\Delta B_t)^2\right)\right|\right)$$
$$< \epsilon(b-a).$$

Hence

$$E\left(\left|f(B_b) - f(B_a) - \int_a^b f'(B_t)\, dB_t - \frac{1}{2}\int_a^b f''(B_t)(dB_t)^2\right|\right) < 3\epsilon(b-a).$$

Thus

$$E\left(f(B_b) - f(B_a) - \int_a^b f'(B_t)\, dB_t - \frac{1}{2}\int_a^b f''(B_t)(dB_t)^2\right) = 0. \quad (3.8)$$

Hence

$$f(B_b) - f(B_a) = \int_a^b f'(B_t)\, dB_t + \frac{1}{2}\int_a^b f''(B_t)(dB_t)^2, \quad (3.9)$$

for almost all $\omega \in \Omega$.

As a consequence, we have given a proof of Itô's formula (3.9).

Definition 3.4. Let $f : \mathbb{R} \to \mathbb{R}$ have continuous second derivative. Suppose the Itô integral $\int_a^b f'(B_t)dBt$ exists and for each $t \in [a,b]$, the Reimann integral $(R)\int_a^b f''(B_t)dt$, $E(f_t)$ and $E\big((R)\int_a^b f_t''(B_t)dt\big)$ exist. We define the differential of $f(B_t)$ as follows:

$$df(B_t) = f'(B_t)\, dB_t + \frac{1}{2}f''(B_t)(dB_t)^2. \quad (3.10)$$

Note that $f(B_{t+\Delta t}) - f(B_t) \neq df(B_t)$. In fact,

$$E\left(|f(B_{t+\Delta t}) - f(B_t) - df(B_t)|\right) = o(\Delta t), \quad (3.11)$$

see inequality (3.5). It is analogous to $df = f'(x)dx$, where $|f(x+\Delta x) - f(x) - df| = o(\Delta x)$.

We can give rigorous definitions of differentials for the Itô integral since we use the Henstock–Kurzweil approach to define the Itô integral.

In the standard stochastic calculus, the differential $df(B_t)$ is defined using equality (3.10) without explicitly pointing out (3.11) where the error of an approximation is $o(\Delta t)$.

As in the deterministic case, by integrating both sides of (3.10),

$$\int_a^a df(B_t) = \int_a^b f'(B_t)\, dB_t + \frac{1}{2}\int_a^b f''(B_t)(dB_t)^2. \quad (3.12)$$

We get Equation (3.9).

Next we consider $V(t, B_t)$, in which $V(t,x)$ is continuously differentiable with respect to t, and twice continuously differentiable with respect to x. Then, by Taylor's formula, we have

$$\begin{aligned} V(t+\Delta t, B_{t+\Delta t}) - V(t, B_t) &= \frac{\partial}{\partial t}V(t,B_t)(\Delta t) + \frac{\partial}{\partial x}V(t,B_t)(\Delta B_t) \\ &\quad + \frac{1}{2}\frac{\partial^2}{\partial x^2}V(t,B_t)(\Delta B_t)^2 + o(\Delta t) + o\left((\Delta B_t)^2\right). \end{aligned}$$

Definition 3.5. We assume that the Itô integral $\int_a^b \frac{\partial}{\partial x} V(t, B_t) dB_t$ exists and for each $t \in [a, b]$, $(R) \int_a^b \frac{\partial}{\partial t} V(t, B_t) dt$ and $(R) \int_a^b \frac{\partial^2}{\partial x^2} V(t, B_t) dt$ exist. Furthermore $E\big((R) \int_a^b \frac{\partial}{\partial t} V(t, B_t) dt\big)$, $E\big((R) \int_a^b \frac{\partial^2}{\partial x^2} V(t, B_t) dt\big)$, $E\left(\frac{\partial}{\partial t} V(t, B_t)\right)$ and $E\big(\frac{\partial^2}{\partial x^2} V(t, B_t)\big)$ exist. We define

$$dV(t, B_t) = \frac{\partial}{\partial t} V(t, B_t) dt + \frac{\partial}{\partial x} V(t, B_t) dB_t + \frac{1}{2} \frac{\partial^2}{\partial x^2} V(t, B_t) (dB_t)^2. \quad (3.13)$$

Similarly, by integrating both sides,

$$\begin{aligned} V(t, B_t) &= \int dV(t, B_t) \\ &= \int \frac{\partial V}{\partial t}(t, B_t) dt + \int \frac{\partial V}{\partial x}(t, B_t) dB_t + \frac{1}{2} \int \frac{\partial^2 V}{\partial x^2}(t, B_t)(dB_t)^2. \end{aligned} \quad (3.14)$$

From the above discussion, when we use differentials $df(B_t)$ and $dV(f, B_t)$ to approximate $f(B_{t+\Delta t}) - f(B_t)$ and $V(t+\Delta t, B_{t+\Delta t}) - V(t, B_t)$, respectively, we have the error term in $L(\Omega)$ to be $o(\Delta t)$. We would like to highlight that the error of an approximation is $o(\Delta t)$, which explains why we can interpret (3.10) and (3.13) in differential forms as (3.9) and (3.14) in integral forms, respectively.

3.2.3 *Differentials for $(dB_t)^2$, $(dt)^2$ and $dtdB_t$*

In this section, we compute some differentials directly which are instrumental for computing the differentials for the Itô integral and the function of a stochastic process presented in later sections. We assume that the conditions in Definitions 3.4 and 3.5 hold.

(i) Let $(dB_t)^2 = (B_{t+\Delta t} - B_t)^2$. Recall that $E(dB_t)^2 = \Delta t$, we have

$$E\left(\left|(dB_t)^2 - \Delta t\right|\right) \leq 2\Delta t.$$

Thus we may use Δt to approximate $(dB_t)^2$ and the error is $o(\Delta t)$. Therefore we define the differential for $(dB_t)^2$ to be Δt.

(ii) Let $dt = (t + \Delta t) - t$. Then $E\left((dt)^2\right) = (dt)^2$. Thus $E\left((dt)^2 - 0\right) = o(\Delta t)$. Therefore we define the differential for $(dt)^2$ to be 0.

(iii) $E\left(|dtdB_t|\right) = |dt| E(|dB_t|) = |dt| \sqrt{\frac{2}{\pi}} \sqrt{dt}$. Thus $E\left(|(dtdB_t) - 0|\right) = o(\Delta t)$. Hence we define the differential for $dtdB_t$ to be 0.

When we do differential computation in this section, we may use $(dt)^2 = 0$, $dtdB_t = 0$ and $(dB_t)^2 = dt$. We remark that in the above we use these substitution in a differential sense and assume the conditions in Definitions 3.4 and 3.5 hold. Hence

$$df(B_t) = f'(B_t)\, dB_t + \frac{1}{2}f''(B_t)(dB_t)^2$$

can be written as

$$df(B_t) = f'(B_t)\, dB_t + \frac{1}{2}f''(B_t)dt,$$

since we can use dt to approximate $(dB_t)^2$ and the error is $o(\Delta t)$, for details, see Theorem 3.6.

Also,

$$dV(t, B_t) = \frac{\partial}{\partial t}V(t, B_t)dt + \frac{\partial}{\partial x}V(t, B_t)dB_t + \frac{1}{2}\frac{\partial^2}{\partial x^2}V(t, B_t)(dB_t)^2$$

can be written as

$$dV(t, B_t) = \frac{\partial}{\partial t}V(t, B_t)dt + \frac{\partial}{\partial x}V(t, B_t)dB_t + \frac{1}{2}\frac{\partial^2}{\partial x^2}V(t, B_t)dt.$$

Example 3.1. Find dB_t^5.

Use

$$df(B_t) = f'(B_t)dB_t + \frac{1}{2}f''(B_t)dt.$$

Let $f(u) = u^5$. So, $f'(u) = 5u^4$ and $f''(u) = 20u^3$. Hence, by the formula above, we get

$$dB_t^5 = 5B_t^4 dB_t + \frac{1}{2}20B_t^3 dt = 5B_t^4 dB_t + 10B_t^3 dt,$$

i.e., we can use $5B_t^4 dB_t + 10B_t^3 dt$ to approximate $B_t^5 - B_{t+dt}^5$ and the error is $o(dt)$. Furthermore the Itô integral $\int_0^t B_s^4 dB_s$ exists. For each $s \in [a, b]$, $(R)\int_0^t B_s^2 ds$, $E(B_s^2)$ and $E\big((R)\int_0^t B_s^2 ds\big)$ exist. Thus

$$B_t^5 = 5\int_0^t B_s^4 dB_s + 10\int_0^t B_s^2 ds.$$

Example 3.2. Find $dt^2 B_t^5$.

Use formula

$$dV(t, B_t) = \frac{\partial}{\partial t}V(t, B_t)dt + \frac{\partial}{\partial x}V(t, B_t)dB_t + \frac{1}{2}\frac{\partial^2}{\partial x^2}V(t, B_t)dt.$$

Let $V(t, x) = t^2x^5$. Then, $\frac{\partial V}{\partial t} = 2tx^5$, $\frac{\partial V}{\partial x} = 5t^2x^4$ and $\frac{\partial^2 V}{\partial x^2} = 20t^2x^3$. Hence, by the formula above, we get

$$\begin{aligned} dt^2 B_t^5 &= 2tB_t^5 dt + 5t^2 B_t^4 dB_t + \frac{1}{2}20t^2 B_t^3 dt \\ &= 2tB_t^5 dt + 5t^2 B_t^4 dB_t + 10t^2 B_t^3 dt. \end{aligned}$$

Thus

$$t^2 B_t^5 = 2\int_0^t sB_s^5 ds + 5\int_0^t s^2 B_s^4 dB_s + 10\int_0^t s^2 B_s^3 ds.$$

3.2.4 *Differential for $f(X_t)$ with $E(\Delta X_t)^2 = o(\Delta t)$*

In this section, we shall discuss differential for $f(X_t)$ where $\{X_t\}$ is a process with $E(\Delta X_t)^2 \le c(\Delta t)$ for small Δt, i.e., $E(\Delta X_t)^2 = o(\Delta t)$. First, we shall give several examples of stochastic processes which have this property.

Example 3.3. Let X_t be a process defined by

$$dX_t = \mu\, dt + \sigma\, dB_t,$$

i.e., $\Delta X_t = \mu\Delta t + \sigma\Delta B_t$.

The process X_t is called Brownian motion with drift μ and variance σ. Here, ΔX_t is denoted by differential dX_t.

Then

$$(dX_t)^2 = (\mu\, dt + \sigma\, dB_t)^2 = \mu^2(dt)^2 + \sigma^2(dB_t)^2 + 2\mu\sigma dt dB_t.$$

By (ii) and (iii) in Subsection 3.2.3, we just need to consider $\sigma^2\left(dB_t\right)^2$. For convenience, we may assume $\left(dX_t\right)^2 = \sigma^2\left(dB_t\right)^2$.

Hence $E\left((dX_t)^2\right) = \sigma^2 E\left((dB_t)^2\right) = \sigma^2\Delta t$.

Example 3.4. Let S_t be an adapted process defined by

$$dS_t = \mu S_t\, dt + \sigma S_t\, dB_t,$$

i.e., $\Delta S_t = \mu S_t\Delta t + \sigma S_t\Delta B_t$.

As in Example 3.3, we may assume

$$(dS_t)^2 = \sigma^2 S_t^2 (dB_t)^2.$$

We have

$$E\left((dS_t)^2\right) = \sigma^2 E\left(S_t^2\right) E\left((dB_t)^2\right).$$

Hence $E\left((dS_t)^2\right) = \sigma^2 E\left(S_t^2\right) dt$.

It is known that the process S_t is a Geometric Brownian Motion with drift μ and variance σ. The process S_t is $S_t = S_0 e^{\hat{\mu}t + \sigma^2 B_t}$, where $\hat{\mu} = \mu - \frac{1}{2}\sigma$, $E(S_t) = S_0 e^{\mu t}$ and $V(S_t) = S_0^2 e^{2\mu t}\left(e^{\sigma^2 t} - 1\right)$. Hence, $E\left(S_t^2\right) = S_0^2 e^{(2\mu+\sigma^2)t}$, see [Calin (2015), pp. 51–52], [Mikosch (1998), pp. 42–43]. Thus $E\left((dS_t)^2\right) = \sigma^2 S_0^2 e^{(2\mu+\sigma^2)t} dt$. Assume that $t \in [a, b]$. Hence $E\left((dS_t)^2\right) \le c\, dt$, for some constant c.

Let $\{X_t\}$ be a process with $E\big((\Delta X_t)^2\big) \le c\Delta t$, where c is a constant and Δt is small. Let $f(x)$ and $V(t, x)$ be the functions given in Subsection 3.2.2. Use the idea in Subsection 3.2.2 and inequality (3.5) and assume that the corresponding conditions in Definitions 3.4 and 3.5 hold, we define the differential for $f(X_t)$ to be

$$df(X_t) = f'(X_t)\,dX_t + \frac{1}{2}f''(X_t)\,(dX_t)^2$$

and

$$dV(t,X_t) = \frac{\partial}{\partial t}V(t,X_t)dt + \frac{\partial}{\partial x}V(t,X_t)dX_t + \frac{1}{2}\frac{\partial^2}{\partial x^2}V(t,X_t)\,(dX_t)^2.$$

Example 3.5. Let $dX_t = \mu dt + \sigma dB_t$. Then

$$\begin{aligned} df(X_t) &= f'(X_t)\,dX_t + \frac{1}{2}f''(X_t)\,(dX_t)^2 \\ &= f'(X_t)\,(\mu dt + \sigma dB_t) + \frac{1}{2}f''(X_t)(\sigma^2 dt) \\ &= \left(\mu f'(X_t) + \frac{\sigma^2}{2}f''(X_t)\right)dt + \sigma f'(X_t)dB_t \end{aligned}$$

and

$$\begin{aligned} dV(t,X_t) &= \frac{\partial}{\partial t}V(t,X_t)dt + \frac{\partial}{\partial x}V(t,X_t)dX_t + \frac{1}{2}\frac{\partial^2}{\partial x^2}V(t,X_t)\,(dX_t)^2 \\ &= \frac{\partial}{\partial t}V(t,X_t)dt + \frac{\partial}{\partial x}V(t,X_t)\,(\mu dt + \sigma dB_t) + \frac{1}{2}\frac{\partial^2}{\partial x^2}V(t,X_t)(\sigma^2 dt) \\ &= \left(\frac{\partial}{\partial t}V(t,X_t) + \mu\frac{\partial}{\partial x}V(t,X_t) + \frac{\sigma^2}{2}\frac{\partial^2}{\partial x^2}V(t,X_t)\right)dt + \sigma\frac{\partial}{\partial x}V(t,X_t)dB_t. \end{aligned}$$

Example 3.6. Let $dS_t = \mu S_t dt + \sigma S_t dB_t$, i.e., $\frac{dS_t}{S_t} = \mu dt + \sigma dB_t$. Let $f(x) = \ln x$. Then

$$\begin{aligned} df(S_t) &= f'(S_t)\,dS_t + \frac{1}{2}f''(S_t)\,(dS_t)^2 \\ d\ln(S_t) &= \frac{1}{S_t}\,dS_t - \frac{1}{2}\frac{1}{S_t^2}\,(\sigma^2 S_t^2 dt) \\ &= \frac{1}{S_t}\,dS_t - \frac{\sigma^2}{2}dt. \end{aligned}$$

Thus

$$\begin{aligned} \int_0^v d\ln(S_t) &= \int_0^v \frac{1}{S_t}\,dS_t - \int_0^v \frac{\sigma^2}{2}dt \\ \ln S_v - \ln S_0 &= \int_0^v (\mu dt + \sigma dB_t) - \frac{\sigma^2}{2}v \\ &= \mu v + \sigma B_v - \frac{\sigma^2}{2}v. \end{aligned}$$

Hence $S_v = S_0 e^{\left(\mu - \frac{\sigma^2}{2}\right)v + \sigma B_v}$.

Example 3.7. Let $G_s = \int_a^s g_t \, dB_t$ be an Itô's integral on $[a, b]$.

By Theorem 3.1 and Definition 3.1, for almost all $t \in [a, b]$, there exists $\delta(t) > 0$ such that

$$E\left(|\Delta G_t - g_t \Delta B_t|^2\right) < \Delta t, \tag{3.15}$$

where $|\Delta t| < \delta(t)$. Using $(c + d)^2 \leq 2c^2 + 2d^2$, we have
$|\Delta G_t|^2 = |(\Delta G_t - g_t \Delta B_t) + (g_t \Delta B_t)|^2 \leq 2\,|(\Delta G_t - g_t \Delta B_t)|^2 + 2\,|(g_t \Delta B_t)|^2$.
Hence

$$E\left(|\Delta G_t|^2\right) < 2\Delta t + 2E(g_t^2)\Delta t = 2\left(1 + E(g_t^2)\right)\Delta t.$$

Hence $\{G_t\}$ is a processes satisfying $E\left(|\Delta G_t|^2\right) \leq c\Delta t$. With this assumption, we define differentials $df(G_t)$ and $V(t, G_t)$ as follows:

$$df(G_t) = f'(G_t)\, dG_t + \frac{1}{2} f''(G_t)\, (dG_t)^2$$

and

$$dV(t, G_t) = \frac{\partial}{\partial t} V(t, G_t) dt + \frac{\partial}{\partial x} V(t, G_t) dG_t + \frac{1}{2} \frac{\partial^2}{\partial x^2} V(t, G_t)\, (dG_t)^2 .$$

Note that $E\big((dG_t)^2\big) = E\big((\int_t^{t+\Delta t} g_s \, dB_s)^2\big) = \int_t^{t+\Delta t} (E(g_s))^2 \, ds$. By (3.15), if we use $g_t \Delta B_t$ to estimate $dG_t = \Delta G_t$, then the error is $o(\Delta t)$ in $L^2(\Omega)$.

Example 3.8. For each $\omega \in \Omega$, let $H_s = \int_a^s h(t, \omega) dt$ be a Riemann integral on $[a, b]$. Then for almost all ω and almost all t, for every $\epsilon > 0$, there exists $\delta(t, \omega) > 0$ such that

$$|\Delta H_t - h(t, \omega)\Delta t| < \epsilon \Delta t$$

whenever $|\Delta t| < \delta(t, \omega)$.

We may assume that $\delta(t, \omega) < 1$, $\epsilon < 1$ for all ω and t. Then

$$\begin{aligned} E\left(|\Delta H_t|^2\right) &= E\left(|\Delta H_t - h(t, \omega)\Delta t + h(t, \omega)(\Delta t)|^2\right) \\ &\leq 2E\left(|\Delta H_t - h(t, \omega)\Delta t|^2\right) + 2E\left(|h^2(t, \omega)(\Delta t)^2|\right) \\ &< 2\left(\epsilon(\Delta t)\right)^2 + 2E\left(h^2(t, \omega)\right)(\Delta t)^2, \text{ whenever } |\Delta t| < \delta(t, \omega) < 1 \\ &< 2\left(1 + E\left(h^2(t, \omega)\right)\right)\Delta t. \end{aligned}$$

Hence $\{H_t\}$ is a process satisfying $E(|\Delta H_t|^2) \leq c\Delta t$, if $E\left(h^2(t, \omega)\right)$ exists for each t. With this assumption, we can define $df(H_t)$ and $dV(t, H_t)$ as in Example 3.7.

Example 3.9. Let $X_t = \int_0^t h_s(\omega) ds + \int_0^t g_s dB_s$ be an Itô process. Suppose $E\left(h_s^2\right)$ exists for all s. Then from Examples 3.7 and 3.8, we define

$$df(X_t) = f'(X_t)\, dX_t + \frac{1}{2} f''(X_t)\, (dX_t)^2$$

and

$$dV(t, X_t) = \frac{\partial}{\partial t} V(t, X_t) dt + \frac{\partial}{\partial x} V(t, X_t) dX_t + \frac{1}{2} \frac{\partial^2}{\partial x^2} V(t, X_t)\, (dX_t)^2 .$$

3.3 Notes and Remarks

In this chapter, belated differentiation and differential of Stochastic processes can be developed since we use the non-uniform Riemann approach to define the Itô integral.

The results and their proofs of belated differentiation can be found in [Toh and Chew (2005)]. The results and their proofs of differential can be found in [Boonpogkrong, Chew and Toh (Under Review)]. Stochastic differential equations using the Henstock–Kurzweil approach can be found in [Tan and Toh (2011–12)].

As a consequence of Section 3.2, we give a proof of Itô's formula, which is more general than the version in Chapter 2. The proof of the Itô's formula in this chapter is easier than the proof in Chapter 2.

In [G. Ladde and S. Ladde (2013), Section 1.3], Itô–Doob type stochastic differentials have been discussed. There are many examples and exercise there.

Chapter 4

Variational Approach to Stochastic Integration

In this chapter, first we give an alternative definition of the Itô integral using a variational approach. In the variational approach, Vitali's cover is not needed. We also use the variational approach to define other stochastic integrals. The variational integral is equivalent to the classical stochastic integral if the integrator is a semimartingale.

4.1 Henstock's Variational Belated Approach to Itô's Stochastic Integral

In this section, we introduce the Henstock Variational Belated (HVB) integral for a stochastic process with respect to the classical Brownian motion.

Let Ω denote the set of all real-valued continuous functions on $[0,1]$ and $\mathbb{R}$ the set of all real numbers.

Let $B = \{B(t,w) : t \in [0,1]\}$ be a classical Brownian motion adapted to $\{\mathcal{F}_t : t \in [0,1]\}$, i.e., B_t is $\mathcal{F}_t$-measurable for each $t \in [0,1]$, where we may take $\mathcal{F}_t$ to be the smallest σ-field generated by $\{B(s,w) : s \leq t\}$.

Let $\mathcal{I}$ be the set of all left-open intervals in $[0,1]$ and $\delta(x)$ be a positive function defined on $[0,1]$. A finite collection of interval-point pairs $D = \{((u_i, v_i], \xi_i)\}_{i=1}^{n}$ is said to be a δ-fine belated partial division of $[0,1]$ if

(i) the intervals $\{(u_i, v_i]\}$ from D are disjoint subintervals from $[0,1]$; and
(ii) for each $i = 1, 2, \ldots, n$, $(u_i, v_i] \subset (\xi_i, \xi_i + \delta(\xi_i))$.

Here we remark that the associated point ξ is to the left of the interval $(u, v]$, hence belated, and that this partial division is *any* collection of disjoint left-open intervals from $[0,1]$, unlike McShane's belated partial division which requires the union of all the intervals must cover almost the entire interval $[0,1]$. Also, there may be some $i \neq j, i, j = 1, 2, \ldots, n$ such that $\xi_i = \xi_j$ although $(u_i, v_i]$ and $(u_j, v_j]$ are disjoint.

We remark that in Chapters 2 and 3, we use $D = \{((\xi_i, v_i), \xi_i)\}_{i=1}^n$, where $u_i = \xi_i$. In fact, in Definition 2.5 of the Itô integrals, we may use $D = \{((u_i, v_i), \xi_i)\}_{i=1}^n$. The integrals defined by using these two different D are equivalent.

Definition 4.1. Let $F(I, \omega)$ be a real-valued function on $\mathcal{I} \times \Omega$. Then F is said to be an *additive function* on $\mathcal{I}$ if for all $I, J \in \mathcal{I}$ which are disjoint and also $I \cup J \in \mathcal{I}$, we have

$$F(I \cup J, \omega) = F(I, \omega) + F(J, \omega)$$

for all $\omega \in \Omega$.

We next present our definition of variational approach to stochastic integral, which we call Henstock Variational Belated integral.

Definition 4.2. Let $\varphi = \{\varphi(t, \omega) : t \in [0, 1]\}$ be a measurable process which is adapted to the standard filtering space $(\Omega, \mathcal{F}, \{\mathcal{F}_t\}, P)$. Then the process φ is said to be *Henstock Variational Belated (or HVB) integrable* if there exists an additive function $F : \mathcal{I} \times \Omega \to \mathbb{R}$ with $F((u, v], \cdot) \in L_2(\Omega)$ for all $(u, v] \in \mathcal{I}$ such that for any $\epsilon > 0$, there exists $\delta(\xi) > 0$ on $[0, 1]$ such that for any δ-fine belated partial division of $[0, 1]$, which we denote by $D = \{((u, v], \xi)\}$, we have

$$E\left|(D)\sum\{\varphi_\xi[B_v - B_u] - F(u, v]\}\right|^2 < \epsilon. \tag{4.1}$$

The integral of φ over the interval $[0, 1]$ with respect to B refers to the functional $F((0, 1], \cdot)$. We shall use the following symbol

$$F((0, 1], \cdot) = (HVB)\int_0^1 \varphi_t dB_t.$$

Remark 4.1. Using the notation of variation, see, e.g., [McShane (1984); Henstock (1991)], we define the variation of a real-valued function f defined on $\mathcal{I} \times \mathbb{R} \times \Omega$, written as $V(f)$, as

$$V(f) = \inf_{\delta(\xi)>0} \sup_D E\left|(D)\sum f((u, v], \xi, \omega)\right|^2,$$

where the supremum is taken over all δ-fine belated partial divisions $D = \{((u,v],\xi)\}$ and the infimum is taken over all positive functions $\delta(\xi)$ on $[0,1]$.

Using the notation of variation in the above remark, Definition 4.2 can be written as:

Definition 4.3. A measurable process $\varphi = \{\varphi(t,\omega) : t \in [0,1]\}$ is said to be HVB integrable on $[0,1]$ if there exists an additive function F defined on $\mathcal{I} \times \Omega$ for which $F((u,v],\cdot) \in L_2(\Omega)$ for any $(u,v] \subset [0,1]$, and

$$V(F - Y) = 0,$$

where

$$Y((u,v],\xi,\omega) = \varphi(\xi,\omega)[B(v,\omega) - B(u,\omega)]$$

or

$$Y((u,v],\xi) = \varphi_\xi[B_v - B_u]$$

for all $\omega \in \Omega$.

Remark 4.2.

(i) The HVB integral of the process $\varphi = \{\varphi(t,\omega) : t \in [0,1]\}$ can be easily seen to be *unique up to variation zero*, i.e., if there is another additive function $G((u,v],\cdot) \in L_2(\Omega)$ which is the HVB integral of φ over $[0,1]$ with respect to B, then

$$V(F - G) = 0.$$

Hereafter, we shall denote the HVB integral of φ with respect to B over $[0,1]$ to be denoted as

$$(HVB)\int_0^1 \varphi_t dB_t.$$

(ii) It is clear from Definition 4.2 that if $\varphi = \{\varphi(t,\omega) : t \in [0,1]\}$ is HVB integrable with respect to B over $[0,1]$, then φ is HVB integrable with respect to B over any subinterval $[a,b] \subset [0,1]$.

(iii) We remark that for the subsequent parts of Section 4.1, we shall use the equality sign loosely to mean equal up to zero variation, that is, two additive functions $F, G : \mathcal{I} \times \Omega \to \mathbb{R}$ are said to be equal $F = G$ if and only if $V(F - G) = 0$.

4.1.1 *Basic properties of the integral*

As mentioned in Remark 4.2, it is clear from Definition 4.2 that if φ is HVB integrable on $[0,1]$ with respect to B, then it is HVB integrable on any $[a,b] \subset [0,1]$. The other properties of HVB integrals are stated below. The proofs of these standard results are easy hence omitted.

Theorem 4.1 (Additivity of Integrals). *If $\varphi^i(t,\omega)$, $i = 1,2,\ldots,n$ is each HVB integrable to $A_i(\omega)$ with respect to B on $[0,1]$, then $\sum_{i=1}^{n} \varphi_i(t,\cdot)$ is HVB integrable on $[0,1]$, i.e.*

$$(HVB)\int_0^1 \sum_{i=1}^{n} \varphi_i(t,\omega)dB(t,\omega) = \sum_{i=1}^{n}(HVB)\int_0^1 \varphi_i(t,\omega)dB(t,\omega).$$

Theorem 4.2 (Integrability over sub-intervals). *Let $0 \le a < c < b \le 1$. If φ is HVB integrable on $[a,c]$ and HVB integrable on $[c,b]$, where $a \le c \le b$, then φ is HVB integrable on $[a,b]$, i.e.*

$$(HVB)\int_a^b \varphi(t,\omega)dB(t,\omega) = \left((HVB)\int_a^c + (HVB)\int_c^b\right)\varphi(t,\omega)dB(t,\omega).$$

4.1.2 *Equivalence theorem*

Recall from Section 1.2 that $\mathcal{L}_2$ denotes the class of all measurable processes $\varphi = \{\varphi(t,\omega) : t \in [0,1]\}$ adapted to the standard filtering space $(\Omega, \mathcal{F}, \{\mathcal{F}_t\}, P)$ such that $\|\varphi\|_{\mathcal{L}_2}$ is finite, where

$$\|\varphi\|_{\mathcal{L}_2}^2 = \int_\Omega \int_0^1 \varphi^2(t,\omega)dtdP = \int_0^1 E[\varphi^2(t,\omega)]dt.$$

In this section we shall show that if $\varphi \in \mathcal{L}_2$, then our definition of HVB integral coincides with the classical Itô integral (where the integrator X is a Brownian motion), see Corollary 4.1. We need the following theorem, for its proof, see [Toh and Chew (1999); Xu and Lee (1992–93)].

Theorem 4.3. *Let $\varphi \in \mathcal{L}_2$. For any $\epsilon > 0$ there exists $\delta(\xi) > 0$ such that for any δ-fine belated partial division $(D) = \{((u,v],\xi)\}$,*

$$\int_{\cup(u,v]} E\left|(D)\sum \left[\varphi(\xi,\omega)\chi_{(u,v]}(t) - \varphi(t,\omega)\right]\right|^2 dt < \epsilon.$$

Corollary 4.1. *If $\varphi \in \mathcal{L}_2$, then φ is HVB integrable on $[0,1]$ and*

$$(HVB)\int_0^1 \varphi(t,\omega)dB(t,\omega) = (I)\int_0^1 \varphi(t,\omega)dB(t,\omega).$$

Proof. Let $\epsilon > 0$ and $D = \{((u, v], \xi)\}$ be a δ-fine partial belated division, where $\delta(\xi) > 0$ is chosen as in Theorem 4.3, so that all conditions there hold. Let D_I denote the set of intervals from the interval-point pairs of D, i.e., $D_I = \{(u, v] : ([u, v], \xi) \in D\}$.

Define the corresponding step process φ_D as follows:

$$\varphi_D(t, \omega) = \begin{cases} 0, & \text{if } t \notin D_I \\ \varphi(\xi, \omega), & \text{if } t \in (u, v] \in D_I. \end{cases}$$

Clearly φ_D is adapted. Hence the value of the classical Itô's integral of φ_D is the partial sum of φ, i.e.

$$(D) \sum \varphi(\xi, \omega)[B(v, \omega) - B(u, \omega)] = (I) \int_0^1 \varphi_D(t, \omega) dB(t, \omega).$$

Thus

$$\begin{aligned}
E&\left|(D) \sum \{\varphi(\xi, \omega)[B(v, \omega) - B(u, \omega)] - (I) \int_u^v \varphi(t, \omega) dB(t, \omega)\}\right|^2 \\
&= E\left|(D) \sum \left\{(I) \int_u^v \varphi_D(t, \omega) dB(t, \omega) - (I) \int_u^v \varphi(t, \omega) dB(t, \omega)\right\}\right|^2 \\
&= E\left|(D) \sum (I) \int_u^v [\varphi_D(t, \omega) - \varphi(t, \omega)] dB(t, \omega)\right|^2 \\
&= E\left|(I) \int_{\cup(u,v]} [\varphi_D(t, \omega) - \varphi(t, \omega)] dB(t, \omega)\right|^2 \\
&= \int_{\cup(u,v]} E|\varphi_D(t, \omega) - \varphi(t, \omega)|^2 dt \\
&= \int_{\cup(u,v]} E|\varphi(\xi, \omega)\chi_{(u,v]}(t) - \varphi(t, \omega)|^2 dt \\
&< \epsilon,
\end{aligned}$$

if $\delta(\xi) > 0$ is chosen to satisfy the conditions of Theorem 4.3. □

We give another proof of Corollary 4.1 as follows: By Theorem 2.7, φ is Itô integrable on $[0, 1]$. Then, by Lemma 2.9 (Henstock's Lemma), φ is HVB integrable on $[0, 1]$.

We remark that Theorem 4.3 and Corollary 4.1 (Equivalence Theorem) hold true even if the integrator is a martingale with deterministic quadratic variation $\langle X \rangle$ which is absolutely continuous (in the classical sense) on $[0, 1]$.

4.2 Weakly Henstock Variational Belated (WHVB) Approach to Stochastic Integral

In this section, let the integrator X be a general stochastic process on a standard filtering space $(\Omega, \mathcal{F}, \{\mathcal{F}_t\}, P)$. We shall study another variational approach which we call *weakly Henstock Variational Belated* approach.

In Subsection 4.1.2, we consider $\delta(\xi) > 0$ on $[0, 1]$, where $\delta(\xi)$ does not depend on $\omega \in \Omega$. From Corollary 4.1 that every Itô integrable function is HVB integrable, we notice that $\delta(\xi)$ can be chosen to be independent of ω, because the quadratic variation of a Brownian motion is deterministic and Fubini's theorem can be applied (which can be seen from transition in Subsection 4.1.2). In this section, integrators are martingales or semi-martingales instead of Brownian motions. The δ-function $\delta(\xi, \omega) > 0$ will depend on $\omega \in \Omega$.

Throughout this section, we need not assume the adaptedness or measurability of the stochastic process X in our construction of the integral. Before we define the weakly Henstock Variational Belated approach to stochastic integral, we shall recall the concept of a stopping time.

Recall that if $T : \Omega \to [0, \infty]$ is a random variable defined on $(\Omega, \mathcal{F}, \{\mathcal{F}_t\}, P)$, then T is a *stopping time* if

$$\{\omega \in \Omega : T(\omega) \leq t\} \in \mathcal{F}_t$$

for each $t \geq 0$.

It follows easily that the following properties of stopping times are true:

Lemma 4.1. *[Yeh (1995), Theorem 3.7, p. 29] Let $T, U : \Omega \to [0, \infty]$ be two stopping times and $c > 0$. Then the followings are stopping times:*

(i) c,
(ii) $T + c$,
(iii) cT $(c > 1)$,
(iv) $\min\{T, U\}$ *and*
(v) $\max\{T, U\}$.

Definition 4.4. Let $\delta : [0, 1] \times \Omega \to (0, \infty)$ be a real-valued function. Then δ is called a *positive locally stopping process* if $\xi + \delta(\xi, \cdot)$ is a stopping time for each $\xi \in [0, 1]$.

We shall next give some examples of positive locally stopping processes.

Example 4.1. The following are some examples of positive locally stopping processes:

(i) $\delta(\xi)$ is deterministic;
(ii) $\delta(\xi, \cdot) > 0$ is a positive stopping time; and
(iii) $\delta(\xi, \cdot) > 0$ is a positive adapted process, that is, δ_ξ is $\mathcal{F}_\xi$-measurable for all $\xi \in [0, 1]$.

Proof. (i) follows from Lemma 4.1(i) since $\xi + \delta(\xi)$ is deterministic for any $\xi > 0$ hence a stopping time.

(ii) also follows from Lemma 4.1(ii) since $\xi + \delta(\xi, \cdot)$ is a stopping time for any stopping time $\delta(\xi, \cdot)$.

We just need to prove (iii), that is, to show that $\xi + \delta(\xi, \cdot)$ is a stopping time for any $\xi \in [0, 1]$. For $t \geq \xi$, we have

$$\{\omega \in \Omega : \xi + \delta(\xi, \cdot) \leq t\} = \{\omega \in \Omega : \delta(\xi, \cdot) \leq t - \xi\} \in \mathcal{F}_\xi \subset \mathcal{F}_t$$

from the adaptedness of δ_ξ. If $t < \xi$, then

$$\begin{aligned}\{\omega \in \Omega : \xi + \delta(\xi, \cdot) \leq t\} &= \{\omega \in \Omega : \delta(\xi, \cdot) \leq t - \xi\} \subset \{\omega \in \Omega : \delta(\xi, \cdot) < 0\} \\ &= \phi \in \mathcal{F}_t.\end{aligned}$$

In either case, $\xi + \delta(\xi, \cdot)$ is a stopping time, hence $\delta(\xi, \omega) > 0$ is a positive locally stopping process. □

4.2.1 *Construction of random intervals and definition of integrals*

Let S and T be two stopping times with $0 \leq S \leq T \leq 1$, i.e. $0 \leq S(\omega) \leq T(\omega) \leq 1$ for each $\omega \in \Omega$. Let $(S, T]$ be a left-open stochastic interval, i.e.

$$(S, T] = \begin{cases} \{(t, \omega) : S(\omega) < t \leq T(\omega)\}, & \text{if } S(\omega) < T(\omega); \\ \{(t, \omega) : S(\omega) = t = T(\omega)\}, & \text{if } S(\omega) = T(\omega). \end{cases}$$

The definition of $(S, T]$ is slightly different from the standard definition in which $S < T$. We include the case $S(\omega) = T(\omega)$ at our own convenience, which enables the clearer presentation of the proof of Theorem 4.12 (see the construction of D_j in the proof of the last part of the theorem). On the other hand, it does not affect Definition 4.6.

We make a simple observation below:

Theorem 4.4. *The stochastic intervals $(S, T] \subset (U, V]$ if and only if*

$$(S(\omega), T(\omega)] \subset (U(\omega), V(\omega)]$$

for all $\omega \in \Omega$.

Remark 4.3. Based on Theorem 4.4 we may loosely say $\xi \in (U, V]$ to mean $\xi \in (U(\omega), V(\omega)]$ for all $\omega \in \Omega$, even though in the strict sense $(U, V]$ refers to a subset of $[0,1] \times \Omega$.

Definition 4.5. Let $\delta(\xi, \omega) > 0$ be a locally stopping process. A finite collection of stochastic interval-point pairs $\{((S_i, T_i], \xi_i)\}_{i=1}^n$ is said to be a δ-fine belated partial stochastic division of $[0,1]$ if

(i) for each i, $(S_i, T_i]$ is a left-open stochastic interval and that for each $\omega \in \Omega$, $(S_i(\omega), T_i(\omega)]$, $i = 1, 2, \ldots, n$, are disjoint left-open subintervals of $[0,1]$ and

(ii) each $((S_i, T_i], \xi_i)$ is δ-fine belated, i.e., for each $\omega \in \Omega$ we have
$$(S_i(\omega), T_i(\omega)] \subset (\xi_i, \xi_i + \delta(\xi_i, \omega)).$$
For the case $S_i(\omega) = T_i(\omega)$, $(S_i(\omega), T_i(\omega)]$ is taken to be $\{T_i(\omega)\}$.

Remark 4.4. Note that for each $\xi \in [0,1]$, $\xi + \delta(\xi, \cdot)$ is a stopping time. Thus for each ξ_i, there exists a stochastic interval $(S_i, T_i]$ with $S_i < T_i$ such that $(S_i(\omega), T_i(\omega)] \subset (\xi_i, \xi_i + \delta(\xi_i, \omega))$ for each $\omega \in \Omega$, namely, $S_i(\omega) = \xi_i$, $T_i(\omega) = \xi_i + \frac{1}{2}\delta(\xi_i, \omega)$.

Definition 4.6. An adapted process $\phi = \{\phi(t, \omega) : t \in [0,1]\}$ in the space $(\Omega, \mathcal{F}, \{\mathcal{F}_t\}, P)$ is said to be *weakly Henstock variational belated*, denoted by WHVB, *integrable* to a process A in $(\Omega, \mathcal{F}, \{\mathcal{F}_t\}, P)$ on $[0,1]$ with respect to a process $X = \{X(t, \omega) : t \in [0,1]\}$ if for every $\epsilon > 0$ there exists a locally stopping process $\delta(\xi, \omega) > 0$ on $[0,1] \times \Omega$ for which
$$E\left(\left|\sum_{i=1}^{n}\{\phi(\xi_i, \omega)(X(T_i(\omega), \omega) - X(S_i(\omega), \omega)) - (A(T_i(\omega), \omega) - A(S_i(\omega), \omega))\}\right|^2\right) < \epsilon,$$
for every $\delta(\omega)$-fine belated partial division $D = \{((S_i, T_i], \xi_i)\}_{i=1}^n$ of $[0,1] \times \Omega$.

For each stopping time U, let X_U denote the mapping from $\Omega \to \mathbb{R}$ such that $X_U(\omega) = X(U(\omega), \omega)$ and for each $\xi \in [0,1]$, let ϕ_ξ denote the mapping $\Omega \to \mathbb{R}$ such that $\phi_\xi(\omega) = \phi(\xi, \omega)$ for each $\omega \in \Omega$. Using the above notation and for succinctness the above expression of WHVB integral may be written as
$$E\left(\left|(D)\sum_{i=1}^{n}\phi_{\xi_i}(X_{T_i} - X_{S_i}) - (A_{T_i} - A_{S_i})\right|\right)^2 < \epsilon.$$

We shall use this notation and assume that X is *cadlag*, see Definition 1.4(iii), throughout this book.

Definition 4.7. Let $\mathcal{I}$ denote the class of all left-open stochastic intervals from $[0,1] \times \Omega$ of the form $(S,T]$, where $0 \leq S \leq T \leq 1$, and let $h((S,T],\xi,\omega)$ be a mapping from $\mathcal{I} \times [0,1] \times \Omega \to \mathbb{R}$. The *variation* of h, denoted by $V(h)$, is defined as

$$V(h) = \inf_{\delta(\xi,\omega)>0} \sup_{D} E \left| (D) \sum h((S,T],\xi,\omega) \right|^2$$

where the supremum is taken over all $\delta(\omega)$-fine belated partial stochastic divisions D of $[0,1] \times \Omega$ and the infimum is taken over all positive locally stopping processes $\delta(\xi,\omega) > 0$.

Then h is said to have *finite quadratic variation* if $V(h)$ is finite.

The above definition of the quadratic variation is different from the classical definition, see, e.g. [Klebaner (2012), p. 220]. Here, we only use belated divisions, which is consistent with our definition of the stochastic integral. Furthermore, we take expectation.

To paraphrase Definition 4.6 in the language of variation in Definition 4.7, it can be reformulated as in Definition 4.8 that follows.

Definition 4.8. An adapted process ϕ in $(\Omega, \mathcal{F}, \{\mathcal{F}_t\}, P)$ is said to be WHVB integrable to a process A in $(\Omega, \mathcal{F}, \{\mathcal{F}_t\}, P)$ on $[0,1]$ if $V(Y-A) = 0$, where

$$Y((S_i,T_i],\xi_i,\omega) = \phi_{\xi_i}(\omega)\,(X_{T_i}(\omega) - X_{S_i}(\omega))$$

and

$$A((S_i,T_i],\xi_i,\omega) = A_{T_i}(\omega) - A_{S_i}(\omega).$$

It is clear that we have:

Lemma 4.2 (Uniqueness Theorem). *If the WHVB integral of an adapted process ϕ with respect to X exists on $[0,1]$, then it is unique up to variation zero.*

If A and B are two WHVB integrals of ϕ with respect to X over $[0,1]$, then we say that A and B are *versions* of the integral. We denote any version of the integral by

$$(WHVB) \int_0^1 \phi_t dX_t.$$

We remark that in Definition 4.2 of HVB integral we use additive functions for the primitives. In Definition 4.6 of WHVB integrals we use processes for the primitive. It is just for convenience. In fact we can use additive functions instead.

Similar to Definition 4.7, we have the following definition, which is different from the classical definition. The value of $V(X)$ without expectation is smaller than the value of the classical version, since the left-hand jumps are not taken into account.

Definition 4.9. A process $X = \{X(t, \omega) : t \in [0, 1]\}$ is said to be of *finite quadratic variation* on $[0, 1]$ if $V(X)$ is finite, where

$$V(X) = \inf_{\delta(\xi,\omega)>0} \sup_{D} E\left|(D)\sum\{X_T - X_S\}\right|^2 .$$

Next we state an important theorem critical to our subsequent results especially on martingales as integrators.

Theorem 4.5 (Optional Sampling Theorem). *[Yeh (1995), pp. 26, 131; Protter (2005), pp. 5, 10] Let $X = \{X(t, \omega) : t \in [0, 1]\}$ be a martingale and S, T be two stopping times such that $0 \le S \le T \le 1$ and let*

$$\mathcal{F}_S = \{A \in \mathcal{F} : A \cap \{\omega \in \Omega : S(\omega) \le t\} \in \mathcal{F}_t \textit{ for all } \ t \in [0, 1]\}.$$

Then

$$E[X_T|\mathcal{F}_S] = X_S$$

for almost all $\omega \in \Omega$.

Example 4.2. Let X be an L^2-martingale on $(\Omega, \mathcal{F}, \{\mathcal{F}_t\}, P)$. Then X is of finite quadratic variation $V(X)$.

Proof. For any sequence of stopping times $\{T_i\}_{i=1}^n$ with $0 \le T_1 \le T_2 \le T_3 \le \cdots \le T_n \le 1$, define

$$\mathcal{F}_{T_i} = \{A \in \mathcal{F} : A \cap \{\omega \in \Omega : T_i(\omega) \le t\} \in \mathcal{F}_t \text{ for all } t \in [0, 1]\} .$$

It could be proved that X_{T_i} is $\mathcal{F}_{T_i}$-measurable for each $i = 1, 2, \ldots, n$ (see [Yeh (1995), Theorem 3.3(2), p. 26]. Also from the Optional Sampling Theorem, see Theorem 4.5, we have that $E(X_{T_{i+1}}|\mathcal{F}_{T_i}) = X_{T_i}$, or equivalently, $E(X_{T_{i+1}} - X_{T_i}|\mathcal{F}_{T_i}) = 0$.

We also have

$$E(X_{T_{i+1}}^2 - X_{T_i}^2) = E(X_{T_{i+1}} - X_{T_i})^2 = E(\langle X\rangle_{T_{i+1}} - \langle X\rangle_{T_i}),$$

where $\langle X \rangle$ is the quadratic variation process associated with the L^2-martingale, see Chapter 1, Definition 1.7.

The proofs of the above equalities are similar to the proofs in Remark 1.2 of Chapter 1.

Thus for any $\delta(\omega)$-fine belated stochastic partial division $D = \{((T_i, T_{i+1}], \xi_i)\}$ of $[0,1] \times \Omega$ we have

$$\begin{aligned} E\left|\sum (X_{T_{i+1}} - X_{T_i})\right|^2 &= E\left[\sum (X_{T_{i+1}} - X_{T_i})^2\right] \\ &\quad + 2E\left[\sum (X_{T_{j+1}} - X_{T_j})(X_{T_{i+1}} - X_{T_i})\right] \\ &= \sum E(X_{T_{i+1}} - X_{T_i})^2 \\ &= \sum \{E\langle X \rangle_{T_{i+1}} - E\langle X \rangle_{T_i}\} \\ &\leq E\langle X \rangle_1 - E\langle X \rangle_0. \end{aligned} \tag{4.2}$$

We remark that the second expression in Equation (4.2) is zero since

$$\begin{aligned} &E\left[\sum_{i<j} (X_{T_{j+1}} - X_{T_j})(X_{T_{i+1}} - X_{T_i})\right] \\ &\quad = \sum_{i<j} E[(X_{T_{j+1}} - X_{T_j})(X_{T_{i+1}} - X_{T_i})] \\ &\quad = \sum_{i<j} E\left\{E[((X_{T_{j+1}} - X_{T_j})(X_{T_{i+1}} - X_{T_i}))\,|\mathcal{F}_{T_i}]\right\} \\ &\quad = \sum_{i<j} E\left\{(X_{T_{j+1}} - X_{T_j})E[(X_{T_{i+1}} - X_{T_i})|\mathcal{F}_{T_i}]\right\} \\ &\quad = 0, \end{aligned}$$

hence completing the proof. □

4.2.2 *Properties of WHVB integral*

We shall next state some standard properties of WHVB integrals without proofs. The proofs are standard in the theory of Henstock integration. However, we shall highlight the following: Let $\delta_i(\xi,\omega), i = 1,2$ be two positive locally stopping processes and let

$$\delta(\xi,\omega) = \min(\delta_1(\xi,\omega), \delta_2(\xi,\omega)).$$

Observe that

$$\xi + \delta(\xi,\omega) = \xi + \min(\delta_1(\xi,\omega), \delta_2(\xi,\omega)) = \min\left(\xi + \delta_1(\xi,\omega), \xi + \delta_2(\xi,\omega)\right)$$

which is a stopping time, hence $\delta(\xi,\omega)$ is a positive locally stopping process. This result is used in the proofs of the following properties.

Theorem 4.6. *If $\phi_i(t,\cdot)$ is WHVB integrable with respect to X on $[0,1]$ for each $i=1,2,\ldots,k$, then $\sum_{i=1}^k \phi_i(t,\cdot)$ is WHVB integrable with respect to X on $[0,1]$ and*

$$(WHVB)\int_0^1 \sum_{i=1}^k \phi_i(t,\cdot)dX(t,\cdot) = \sum_{i=1}^k (WHVB)\int_0^1 \phi_i(t,\cdot)dX(t,\cdot).$$

It is easy to see from the definition that if ϕ is WHVB integrable with respect to X on $[0,1]$, then it is WHVB integrable with respect to X on each $[a,b]\subset[0,1]$. We state the next property.

Theorem 4.7. *If ϕ is WHVB integrable on $[0,c]$ and WHVB integrable on $[c,1]$, where $0\le c\le 1$, then ϕ is $WHVB$ integrable on $[0,1]$. Furthermore,*

$$(WHVB)\int_0^1 \phi_t dX_t = (WHVB)\int_0^c \phi_t dX_t + (WHVB)\int_c^1 \phi_t dX_t.$$

Theorem 4.8. *If ϕ is WHVB integrable with respect to X and Y, then ϕ is WHVB integrable with respect to $X\pm Y$ and moreover,*

$$(WHVB)\int_0^1 \phi_t d(X_t\pm Y_t) = (WHVB)\int_0^1 \phi_t dX_t \pm (WHVB)\int_0^1 \phi_t dY_t.$$

4.2.3 *Integration by parts and related results*

It is well-known in the classical theory of Riemann–Stieltjes integration that when one of the functions f and g is continuous and the other of bounded variation on $[a,b]$, we have the integration-by-parts formula

$$(RS)\int_a^b f dg + (RS)\int_a^b g df = f(b)g(b) - f(a)g(a)$$

where f and g are deterministic functions on $[a,b]$. Consequently, by putting $f=g$, we have

$$(RS)\int_a^b f df = \frac{1}{2}f^2(b) - \frac{1}{2}f^2(a).$$

In this section we shall show that if f and g are general stochastic processes with paths of unbounded variation, in particular, if X is a martingale, the above formulae may not be true, see Equation (4.4).

We remark that we may replace the δ-fine division $\{((U,V],\xi)\}$ by $\{((\xi,V],\xi)\}$ in Definition 4.6. These two versions are equivalent, which can be seen from the proofs of Theorem 4.13. We shall hence use $\{((\xi, V], \xi)\}$ to simplify the proofs of results in this section.

Theorem 4.9. *Let ϕ and X be adapted processes defined on the standard filtering space $(\Omega, \mathcal{F}, \mathcal{F}_t\}, P)$. Assume that ϕ is WHVB integrable with respect to X on $[0,1]$. Suppose that there exists a process H such that $V(\Delta\phi\Delta X - \Delta H) = 0$, where $\Delta Y(U,V] = Y_V - Y_U$. Then X is WHVB integrable with respect to ϕ on $[0,1]$. Furthermore,*

$$(WHVB)\int_0^1 \phi_t dX_t + (WHVB)\int_0^1 X_t d\phi_t = \phi_1 X_1 - \phi_0 X_0 - (H_1 - H_0).$$

Before we give the proof, we remark that a similar result has been proved for deterministic functions, see, e.g., [Henstock (1988), p. 64]. The following proof is similar to the deterministic case.

Proof. Let $F_u = (WHVB)\int_0^u \phi_t dX_t$ and define

$$G_u = \phi_u X_u - (WHVB)\int_0^u \phi_t dX_t - H_u,$$

so that

$$G_1 - G_0 = \phi_1 X_1 - \phi_0 X_0 - (WHVB)\int_0^1 \phi_t dX_t - (H_1 - H_0).$$

Given $\epsilon > 0$ there exists a positive locally stopping process $\delta_1(\xi, \omega) > 0$ such that for any $\delta_1(\omega)$-fine belated partial stochastic division $D_1 = \{((\xi_i, V_i], \xi_i)\}_{i=1}^{m(1)}$ of $[0,1] \times \Omega$, we have

$$E\left|(D_1)\sum_{i=1}^{m(1)} \{\phi_{\xi_i}[X_{V_i} - X_{\xi_i}] - (F_{V_i} - F_{\xi_i})\}\right|^2 < \frac{\epsilon}{2}.$$

In view of

$$V(\Delta\phi\Delta X - \Delta H) = 0$$

we can choose a positive locally stopping process $\delta_2(\xi, \omega) > 0$ which satisfies the condition that for all $\delta_2(\omega)$-fine belated partial stochastic division $D_2 = \{((\xi_i, V_i], \xi_i)\}_{i=1}^{m(2)}$ we have

$$E\left|(D_2)\sum_{i=1}^{m(2)} \{[\phi_{V_i} - \phi_{\xi_i}][X_{V_i} - X_{\xi_i}] - (H_{V_i} - H_{\xi_i})\}\right|^2 < \frac{\epsilon}{2}.$$

Consider the positive locally stopping process $\delta(\xi, \omega) = \min(\delta_1(\xi, \omega), \delta_2(\xi, \omega))$.

Let $D = \{((\xi_i, V_i], \xi_i)\}_{i=1}^m$ be any $\delta(\omega)$-fine belated partial stochastic division of $[0,1] \times \Omega$. For succinctness we may ignore the index i if the division D used is clear.

Using the following identity

$$X_\xi[\phi_V - \phi_\xi] + \phi_\xi[X_V - X_\xi] = \phi_V X_V - \phi_\xi X_\xi - [\phi_V - \phi_\xi][X_V - X_\xi]$$

and the definition of G_V we get

$$\begin{aligned}
&E\Big(\Big|(D)\sum\{X_\xi[\phi_V - \phi_\xi] - (G_V - G_\xi)\}\Big|\Big)^2 \\
&= E\Big(\Big|(D)\sum\{X_\xi[\phi_V - \phi_\xi] - \{\phi_V X_V - F_V - H_V - \phi_\xi X_\xi + F_\xi + H_\xi\}\}\Big|\Big)^2 \\
&= E\Big(\Big|(D)\sum\{-\phi_\xi[X_V - X_\xi] - (F_\xi - F_V) - [\phi_V - \phi_\xi][X_V - X_\xi] \\
&\quad - (-H_V + H_\xi)\}\Big|\Big)^2 \\
&\leq 2E\left(\Big|(D)\sum\{\phi_\xi[X_V - X_\xi] - (F_V - F_\xi)\}\Big|\right)^2 \\
&\quad + 2E(|(D)\sum\{-[\phi_V - \phi_\xi][X_V - X_\xi] - (-H_V + H_u)\}\,|)^2 \\
&< 2\epsilon + 2\alpha,
\end{aligned}$$

where

$$\begin{aligned}
\alpha &= E\left(\Big|(D)\sum\{-[\phi_V - \phi_\xi][X_V - X_\xi] - (-H_V + H_\xi)\}\Big|\right)^2 \\
&= E\left(\Big|(D)\sum\{[\phi_V - \phi_\xi][X_V - X_\xi] - (H_V - H_\xi)\}\Big|\right)^2 \\
&< \epsilon
\end{aligned}$$

thereby completing our proof. □

An immediate consequence of Theorem 4.9 by putting $H_t = 0$ yields the following integration-by-parts formula:

Corollary 4.2. *Let ϕ and X be adapted processes defined on the standard filtering space $(\Omega, \mathcal{F}, \{\mathcal{F}_t\}, P)$. If ϕ is WHVB integrable with respect to X on $[0,1]$ and $V(\Delta\phi\Delta X) = 0$, then X is WHVB integrable with respect to ϕ on $[0,1]$. Furthermore,*

$$(WHVB)\int_0^1 \phi_t dX_t + (WHVB)\int_0^1 X_t d\phi_t = \phi_1 X_1 - \phi_0 X_0.$$

By Corollary 4.2, we have the following theorem. However, we shall give a different proof.

Theorem 4.10. *Let X be an adapted process defined on $(\Omega, \mathcal{F}, \{\mathcal{F}_t\}, P)$ such that*

$$V\left([\Delta X]^2\right) = 0.$$

Then X is WHVB integrable with respect to X and

$$(WHVB) \int_0^1 X_t dX_t = \frac{1}{2}X_1^2 - \frac{1}{2}X_0^2.$$

Proof. Given $\epsilon > 0$, there exists a positive locally stopping process $\delta(\xi, \omega) > 0$ such that for all $\delta(\omega)$-fine belated partial stochastic division $D = \{((\xi, V], \xi)\}$,

$$E\left(\left|(D)\sum[X_V - X_\xi]^2\right|\right)^2 < 4\epsilon.$$

For any $\delta(\omega)$-fine belated partial stochastic division $D = \{(\xi, V], \xi\}$,

$$\begin{aligned}
&E\left|(D)\sum\{X_\xi(X_V - X_\xi) - \frac{1}{2}(X_V^2 - X_\xi^2)\}\right|^2 \\
&\quad = E\left|(D)\sum\{X_\xi(X_V - X_\xi) - \frac{1}{2}[X_V - X_\xi][X_V + X_\xi]\}\right|^2 \\
&\quad = \frac{1}{4}E\left|(D)\sum\{(X_V - X_\xi)[X_\xi - X_V]\}\right|^2 \\
&\quad < \epsilon,
\end{aligned}$$

thereby completing our proof. □

By Theorem 4.9, we have the following theorem. However, we shall give a different proof.

Theorem 4.11. *Let X be an adapted process defined on $(\Omega, \mathcal{F}, \{\mathcal{F}_t\}, P)$ such that there exists a process G with $V((\Delta X)^2 - \Delta G) = 0$. Then X is WHVB integrable with respect to itself, and*

$$(WHVB) \int_0^1 X_t dX_t = \frac{1}{2}X_1^2 - \frac{1}{2}X_0^2 - \frac{1}{2}(G_1 - G_0).$$

Proof. Given $\epsilon > 0$, let $\delta(\xi, \omega) > 0$ be a positive locally stopping process such that for any $\delta(\omega)$-fine belated partial stochastic division $D = \{((\xi, V], \xi)\}$ of $[0, 1] \times \Omega$ we have

$$E\left|(D)\sum\{[X_V - X_\xi]^2 - (G_V - G_\xi)]\}\right|^2 < \epsilon.$$

Then

$$E\left|(D)\sum\{X_\xi[X_V - X_\xi] - [\frac{1}{2}(X_V^2 - X_\xi^2) - \frac{1}{2}(G_V - G_\xi)]\right|^2$$
$$= E\left|(D)\frac{1}{2}\sum\{2X_\xi(X_V - X_\xi) - (X_V - X_\xi)(X_V + X_\xi) + (G_V - G_\xi)]\}\right|^2$$
$$= \frac{1}{4}E\left|(D)\sum\{-(X_V - X_\xi)^2 + (G_V - G_\xi)\}\right|^2$$
$$= \frac{1}{4}E\left|(D)\sum\{(X_V - X_\xi)^2 - (G_V - G_\xi)\}\right|^2$$
$$< \frac{1}{4}\epsilon,$$

thereby completing the proof. Note that the second equality is the same as in the proof of previous theorem. □

Remark 4.5. Let X be a cadlag L_2-martingale and $\langle X\rangle$ be the associated quadratic variation process. Recall that every martingale is adapted, see Chapter 1, Definition 1.3. It can be proved that

$$V\left((\Delta X)^2 - \Delta\langle X\rangle\right) = 0. \tag{4.3}$$

We remark that to prove Equation (4.3) we need Lemma 4.3.

Lemma 4.3. *Let $\varphi = \{\varphi(t,\omega) : t \in [0,1]\}$ be an adapted cadlag process on $[0,1]$, that is, $\varphi(\cdot,\omega)$ is right-continuous on $[0,1]$ for all $\omega \in \Omega$. Then for each $\omega \in \Omega$, given any $\epsilon > 0$, there exists a positive locally stopping process $\delta(\xi,\omega) > 0$ such that*

$$|\varphi(u(\omega),\omega) - \varphi(\xi,\omega)| < \epsilon$$

for all $u(\omega)$ with $0 < u(\omega) - \xi < \delta(\xi,\omega)$.

Proof. Let $\epsilon > 0$ be given and let $\xi \in [0,1]$. Define for any $(\xi,\omega) \in [0,1]\times\Omega$ the process δ as

$$\delta(\xi,\omega) = \inf\{u \in [0,\infty) : |\varphi(u+\xi,\omega) - \varphi(\xi,\omega)| \geq \epsilon\}.$$

We claim that $\delta(\xi,\omega)$ is a positive stopping process. This statement is true since for $t \geq 0$,

$$\begin{aligned}\{\omega\in\Omega : \xi + \delta(\xi,\omega) \leq t\} &= \{\omega\in\Omega : \delta(\xi,\omega) \leq t - \xi\}\\ &\subseteq \bigcup_{u\in[0,t-\xi]} \{\omega\in\Omega : |\varphi(\xi+u,\omega) - \varphi(\xi,\omega)| \geq \epsilon\}\\ &\in \mathcal{F}_{\xi+(t-\xi)} = \mathcal{F}_t\end{aligned}$$

so that $\xi + \delta(\xi,\cdot)$ is a stopping time for all $\xi \in [0,1]$. Hence for any u with $u - \xi < \delta(\xi,\omega)$ we have $|\varphi(\xi+u,\omega) - \varphi(\xi,\omega)| < \epsilon$. □

Proof of Remark 4.5. Given $\epsilon > 0$, by Lemma 4.3, there exists a positive locally stopping process $\delta(\xi, \omega) > 0$ such that whenever $0 < v(\omega) - \xi < \delta(\xi, \omega)$ we have

$$|\langle X\rangle(v(\omega), \omega) - \langle X\rangle(\xi, \omega)| < \epsilon \text{ and } |X(v(\omega), \omega) - X(\xi, \omega)| < \sqrt{\epsilon}.$$

Choose a δ-fine belated partial stochastic division $D = \{((U, V], \xi)\}$. Then

$$\begin{aligned} E\Big|\sum\left((X_V - X_U)^2 - (\langle X\rangle_V - \langle X\rangle_U)\right)\Big|^2 &\\ = E\Big\{\sum\left[(X_V - X_U)^2 - (\langle X\rangle_V - \langle X\rangle_U)\right]^2\Big\} &\\ \leq \sum E(X_V - X_U)^4 + \sum E(\langle X\rangle_V - \langle X\rangle_U)^2 &\\ + 2\sum E\left[(X_V - X_U)^2(\langle X\rangle_V - \langle X\rangle_U)\right] &\\ < \epsilon \sum E(X_V - X_U)^2 + 3\epsilon \sum E(\langle X\rangle_V - \langle X\rangle_U) &\\ \leq 4\epsilon\left(\langle X\rangle_1 - \langle X\rangle_0\right) & \end{aligned}$$

thereby completing our proof. □

Example 4.3. By Theorem 4.11 and Remark 4.5 we have

$$(WHVB)\int_0^1 X_t dX_t = \frac{1}{2}X_1^2 - \frac{1}{2}X_0^2 - \frac{1}{2}(\langle X\rangle_1 - \langle X\rangle_0).$$

We remark that for the classical stochastic integral, in the above equality, $\int_0^1 X_{t-} dX_t$ replaces $(WHVB)\int_0^1 X_t dX_t$, see, e.g. [Klebaner (2012), Theorem 8.6, p. 221], and the definition of $\int_0^1 X_{t-} dX_t$ uses the standard partition, which is different from the stochastic integral for L^2-martingale.

As a special example, if X is a Brownian motion, then $\langle X\rangle_t \equiv t$. So

$$(WHVB)\int_0^1 X_t dX_t = \frac{1}{2}X_1^2 - \frac{1}{2}X_0^2 - \frac{1}{2}. \tag{4.4}$$

In fact, we see easily that the choice of $\delta(\xi) > 0$ can be made independent of Ω (i.e. deterministic) when we repeat the proof of Theorem 4.11 and replace the integrator X by a Brownian motion.

4.3 Equivalence of WHVB Integral and Classical Stochastic Integral

In this section we shall show that if the integrator X is a semimartingale and the integrand is classical stochastic integrable with respect to X, then the WHVB integral is equivalent to the classical stochastic integral. Our proof of equivalence theorem is first established for a martingale as integrator, and is then generalized to local martingales and then semimartingales.

We remark that an equivalence theorem has been proved in [Protter (1979)], where Lipschitz conditions for the integrators are assumed.

4.3.1 *Convergence theorem and equivalence theorem*

Let the integrator X be an L^2-martingale. Recall that we always assume that X is *cadlag*, see Definition 1.4, Chapter 1. In this section, we also assume that $X(0,\omega) = 0$ a.s.

Now we shall introduce the concept of predictable sets. The family of subsets of $[0,1] \times \Omega$ containing all sets of the form $\{0\} \times F_0$ and $(s,t] \times F$, where $F_0 \in \mathcal{F}_0$ and $F \in \mathcal{F}_s$ for all $s < t$ in $[0,1]$ is called the class of predictable rectangles and we denote it by $\Re$. The σ-field $\Im$ of subsets of $[0,1] \times \Omega$ generated by $\Re$ is called the $\{\mathcal{F}_t\}$-predictable σ-field and sets in $\Im$ are called the $\{\mathcal{F}_t\}$-predictable sets. A process $u = \{u_t(\omega) : t \in [0,1]\}$ is called $\{\mathcal{F}_t\}$-predictable if it is $\Im$-measurable, see [Chung and Williams (1990), p. 25] and [Weizsacker and Winkler (1990), p. 112]. It is known that the $\{\mathcal{F}_t\}$-predictable σ-field $\Im$ can also be generated by all left-continuous $\{\mathcal{F}_t\}$-adapted processes, see [Yeh (1995), p. 19] and [Klebaner (2012), p. 243].

Let $\langle X \rangle$ be the quadratic variation process associated with the L^2-martingale X and $\mathcal{L}_2^{\langle X \rangle}$ denote the class of all measurable, adapted and $\{\mathcal{F}_t\}$-predictable processes $\varphi_t = \{\varphi(t,\omega) : t \in [0,1]\}$ such that

$$\|\varphi\|_{\mathcal{L}_2}^2 = \int_\Omega \int_0^1 \varphi^2(t,\omega)d\langle X \rangle_t dP < \infty,$$

where, for each ω, $\int_0^1 \varphi^2(t,\omega)d\langle X \rangle_t$ is the Lebesgue–Stieltjes integral, see [Yeh (1995), pp. 211, 215].

The Föllmer–Doleans measure μ_X is defined on $\Im$ by

$$\mu_X((s,t] \times F) = E\left(1_F(\langle X \rangle_t - \langle X \rangle_s)\right) = \int_F \left(\langle X \rangle_t(\omega) - \langle X \rangle_s(\omega)\right) dP.$$

Observe that

$$\mu_X((s,t] \times F) = E\left(1_F(X_t^2 - X_s^2)\right).$$

Lemma 4.4. *Let X be a cadlag L^2-martingale. Then the function $f(t) \equiv E\langle X \rangle_t$ is right-continuous on $[0,1]$.*

Proof. $\langle X \rangle_t$ is a right-continuous increasing process for almost all $\omega \in \Omega$ since X is cadlag. By the Dominated Convergence Theorem, the function $f(t) \equiv E\langle X \rangle_t$ is right-continuous on $[0,1]$. □

Lemma 4.5. *Let $f(\omega)$ be a bounded random variable on $(\Omega, \mathcal{F}, P)$ with $|f(\omega)| \leq M$ for all $\omega \in \Omega$. Let $s \in [0,1]$ be fixed. Suppose $u_t(\omega) = f(\omega)$ if $t = s$ and $u_t(\omega) = 0$ if $t \neq s$. Then u is both WHVB and classical stochastic integrable with respect to X on $[0,1]$. Furthermore the integrals agree and equal to zero.*

Proof. First we shall prove that u is WHVB integrable on $[0,1]$ and its integral is zero. Recall that $\langle X \rangle$ denote the quadratic variation process of X. Note that $E\langle X \rangle_t$ is right-continuous in $t \in [0,1]$ by Lemma 4.4. Let $s \in [0,1]$ be fixed. Then there exists $\gamma > 0$ such that $E\langle X \rangle_y - E\langle X \rangle_s < \epsilon$ whenever $0 < y - s \leq \gamma$. Define $\delta(s,\omega) = \gamma$ for all $\omega \in \Omega$. Now if $t \neq s$, define $\delta(t,\omega) = \frac{|t-s|}{2}$ for all $\omega \in \Omega$. Let $D = \{((S_i, T_i], \xi_i)\}_{i=1}^n$ be a δ-fine belated partial stochastic division of $[0,1]$. We may assume there exists j such that $\xi_j = s$. Otherwise it is trivial. So

$$\begin{aligned}
E\left(\left|\sum_{i=1}^{n}\{u_{\xi_i}(X_{T_i} - X_{S_i}) - 0\}\right|^2\right) &= E\left(|u_s(X_{T_j} - X_{S_j})|^2\right) \\
&\leq M^2 E\left((X_{T_j} - X_{S_j})^2\right) \\
&= M^2 E\left(X_{T_j}^2 - X_{S_j}^2\right).
\end{aligned}$$

In the above we have used the fact that X is a martingale and T_j and S_j are bounded stopping times, see [Revuz and Yor (1994), p. 66]. Then

$$\begin{aligned}
E(X_{T_j}^2 - X_{S_j}^2) &\leq |E(X_{T_j}^2) - E(X_s^2)| + |E(X_s^2) - E(X_{S_j}^2)| \\
&= |E(\langle X \rangle_{T_j}) - E(\langle X \rangle_s)| + |E(\langle X \rangle_{S_j}) - E(\langle X \rangle_s)| \\
&\leq |E\langle X \rangle_{\gamma+s} - E\langle X \rangle_s| + |E\langle X \rangle_{\gamma+s} - E\langle X \rangle_s| \\
&< 2\epsilon.
\end{aligned}$$

Note that in the above we have used the fact that X is a right-continuous martingale; $s \leq S_j(\omega), T_j(\omega) \leq \gamma + s$ for each $\omega \in \Omega$ and $\langle X \rangle_t$ is increasing. Finally

$$E\left(\left|\sum_{i=1}^{n}\{u_{\xi_i}(X_{T_i} - X_{S_i})\}\right|^2\right) < 2M^2\epsilon.$$

Hence u is WHVB integrable on $[0,1]$ and its integral is zero. On the other hand, from the classical theory of stochastic integration, it is clear that u is classical stochastic integrable on $[0,1]$ and its integral is zero. □

Lemma 4.6. *Let u be an adapted simple process on the standard filtering space $(\Omega, \mathcal{F}, \{\mathcal{F}_t\}, P)$. Then it is both classical stochastic and WHVB integrable on $[0,1]$. Furthermore, the integrals agree.*

Proof. Let u be an adapted simple process in $\mathcal{L}_0$, i.e.,

$$u_t = f_0 1_{\{0\}}(t) + \sum_{j=0}^{m-1} f_j 1_{(s_j, s_{j+1}]}(t)$$

where $0 = s_0 < s_1 < s_2 < \cdots < s_m = 1$, each f_j is $\mathcal{F}_{s_j}$-measurable and there exists $M > 0$ such that $|f_j(\omega)| \leq M$ for all $\omega \in \Omega$ and all j. By Lemma 4.5 and the fact that there are only finite number of s_j, we may assume that $u_{s_j} = 0$ for all j. Now we shall define $\delta(\xi, \omega)$. If $\xi \neq s_j$ for all j, assuming that $\xi \in (s_k, s_{k+1})$, define $\delta(\xi, \omega) = \frac{1}{2}\min(\xi - s_k, s_{k+1} - \xi)$ for all $\omega \in \Omega$; if $\xi = s_k$ for some k, define $\delta(\xi, \omega) = \frac{1}{2}\min(s_k - s_{k-1}, s_{k+1} - s_k)$ for all $\omega \in \Omega$. Now let $D = \{((S_i, T_i], \xi_i)\}_{i=1}^n$ be a δ-fine belated partial stochastic division of $[0,1]$, and $D_1 = \{((S_i, T_i], \xi_i) \in D : \xi_i \neq s_j \text{ for all } j\}$. Let $(I)\int u_t dX_t$ denote the classical stochastic integral. Then

$$\begin{aligned} &E\left(\left|\sum_{i=1}^n \{u_{\xi_i}(X_{T_i} - X_{S_i}) - (I)\int_{S_i}^{T_i} u_t dX_t\}\right|^2\right) \\ &= E\left(\left|(D_1)\sum_i \{u_{\xi_i}(X_{T_i} - X_{S_i}) - (I)\int_{S_i}^{T_i} u_t dX_t\}\right|^2\right) \\ &= E\left(\left|(D_1)\sum_i \{u_{\xi_i}(X_{T_i} - X_{S_i}) - u_{\xi_i}(X_{T_i} - X_{S_i})\}\right|^2\right) \\ &= 0. \end{aligned}$$

Hence u is WHVB integrable on $[0,1]$ and the integral agrees with the value of the classical stochastic integral. □

Lemma 4.7. *Let $\mathcal{O} \subset [0,1] \times \Omega$. Suppose that $\mathcal{O}$ is predictable with $\mu_X(\mathcal{O}) = 0$. Then $1_{\mathcal{O}}$ is WHVB integrable to zero on $(0,1]$.*

Proof. Suppose $\mathcal{O}$ is predictable with $\mu_X(\mathcal{O}) = 0$. Then given $\epsilon > 0$, there exists a countable collection $\{(s_j, t_j] \times F_j\}_{j=1}^\infty$ of predictable rectangles such that $\mathcal{O} \subset \bigcup_{j=1}^\infty ((s_j, t_j] \times F_j)$ with $\sum_{j=1}^\infty \mu_X((s_j, t_j] \times F_j) < \epsilon$. We need only to consider (s_j, t_j) instead of $(s_j, t_j]$ by Lemma 4.5. Now we shall define $\delta(\xi, \omega)$ on $[0,1] \times \Omega$. If $(\xi, \omega) \in \mathcal{O}$, then $(\xi, \omega) \in (s_j, t_j) \times F_j$ for some j. Choose j to be the smallest of such value. Define $\delta(\xi, \omega)$ such that $\xi + \delta(\xi,\omega) < t_j$ if $\xi \in (s_j, t_j)$ and $\omega \in F_j$. On the other hand, if $(\xi,\omega) \notin \mathcal{O}$, let $\delta(\xi,\omega)$ to be any positive value. Hence $\delta(\xi,\omega)$ is a locally stopping process.

Let $D = \{((S_i, T_i], \xi_i)\}_{i=1}^n$ be a δ-fine belated partial stochastic division of $[0,1]$. We may assume that each ξ_i is in $\bigcup_{j=1}^{\infty}(s_j, t_j)$. Furthermore, for convenience, we may assume that for each (s_j, t_j), there exists at most one $\xi_i \in (s_j, t_j)$. We also assume that $i = j$. In other words, $\xi_i \in (s_i, t_i)$ for each $i = 1, 2, \ldots, n$. Now,

$$
\begin{aligned}
E\left(\left|\sum_{i=1}^{n} 1_{\mathcal{O}}(\xi_i,\omega)(X_{T_i} - X_{S_i}) - 0\right|^2\right) &= E\left|\sum_{i=1}^{n} 1_{F_i}(X_{T_i} - X_{S_i})\right|^2 \\
&= E\left(\sum_{i=1}^{n} 1_{F_i}(X_{T_i} - X_{S_i})^2\right) \\
&= E\left(\sum_{i=1}^{n} 1_{F_i}(X_{T_i}^2 - X_{S_i}^2)\right) \\
&\le \sum_{i=1}^{n} \mu_X\left((s_i, t_i] \times F_i\right) \\
&< \epsilon
\end{aligned}
$$

thereby completing the proof. □

Now we shall prove the following two lemmas before we prove Theorem 4.12 (Variational Convergence Theorem).

Lemma 4.8. *Let u and $u^{(n)}$, $n = 1, 2, \ldots$, be adapted processes on $(\Omega, \mathcal{F}, \{\mathcal{F}_t\}, P)$. Suppose that $u^{(n)}$ converges to u everywhere in $[0,1] \times \Omega$. Then for every $\epsilon > 0$, there exists an adapted process $n(\xi,\omega)$ with positive integral values such that*

$$\left|u^{n(\xi,\omega)}(\xi,\omega) - u(\xi,\omega)\right| < \epsilon$$

for each $(\xi,\omega) \in [0,1] \times \Omega$.

Proof. By given condition, for each $(\xi,\omega) \in [0,1] \times \Omega$, there exists a smallest positive integer $n(\xi,\omega)$ such that

$$\left|u^{n(\xi,\omega)}(\xi,\omega) - u(\xi,\omega)\right| < \epsilon.$$

We shall show that $n(\xi,\omega)$ is adapted. Note that for fixed $\xi \in [0,1]$ and any positive integer N,

$$\begin{aligned}&\{\omega \in \Omega : n(\xi,\omega) \leq N\}\\&= \left\{\omega \in \Omega : |u^{(m)}(\xi,\omega) - u(\xi,\omega)| < \epsilon \text{ for some } m \leq N\right\}\\&= \bigcup_{m=1}^{N} \left\{\omega \in \Omega : |u^{(m)}(\xi,\omega) - u(\xi,\omega)| < \epsilon\right\}\end{aligned}$$

hence $n(\xi,\omega)$ is adapted. □

Lemma 4.9. *Let $n(\xi,\omega)$ defined on $[0,1] \times \Omega$ be an adapted process with positive integral values. Suppose that $\delta^k(\xi,\omega), k = 1,2,\ldots,$ are positive locally stopping processes. Then the process $\delta(\xi,\omega) = \delta^{n(\xi,\omega)}(\xi,\omega)$ is a locally stopping time.*

Proof. For any fixed $\xi \in [0,1]$ and any fixed $t \geq 0$,

$$\begin{aligned}&\{\omega \in \Omega : \xi + \delta(\xi,\omega) \leq t\}\\&= \bigcup_{k=1}^{\infty} \left(\{\omega : \xi + \delta^k(\xi,\omega) \leq t\} \cap \{\omega : n(\xi,\omega) = k\}\right) \in \mathcal{F}_\xi.\end{aligned}$$

Hence $\delta(\xi,\omega)$ is a locally stopping time. □

Definition 4.10. Let A and $A^{(n)}$, $n = 1,2,\ldots,$ be processes on the standard filtering space $(\Omega,\mathcal{F},\{\mathcal{F}_t\},P)$. We say that $A^{(n)}$ *variationally converges* to A if for every $\epsilon > 0$, there exists a positive integer N such that for any finite collection of disjoint stochastic intervals $\{(S_i,T_i]\}_{i=1}^{q}$, we have

$$E\left(\left|\sum_{i=1}^{q}\{(A_{T_i}^{(n)} - A_{S_i}^{(n)}) - (A_{T_i} - A_{S_i})\}\right|^2\right) < \epsilon$$

whenever $n \geq N$.

For variational convergence, see also Lemma 1.3, Chapter 1, and Definition 2.10, Chapter 2.

Example 4.4. Let u and $u^{(n)}$, $n = 1,2,\ldots,$ be classical stochastic integrable on $[0,1]$ with

$$\lim_{n\to\infty} E\left(\int_0^1 (u^{(n)} - u)_t^2 d\langle X\rangle_t\right) = 0.$$

Suppose that $A(s) = (I)\int_0^s u_t dX_t$ and $A^{(n)}(s) = (I)\int_0^s u_t^{(n)} dX_t$ where $(I)\int$ indicates the classical stochastic integral. We shall prove that $A^{(n)}$ variationally converges to A.

Proof. Let $\{(S_i, T_i]\}_{i=1}^n$ be any finite collection of disjoint stochastic intervals. By the orthogonal increment of martingales and the isometric property of the classical stochastic integral,

$$\begin{aligned} E&\left(\left|\sum_{i=1}^{n}\{(A_{T_i}^{(n)} - A_{S_i}^{(n)}) - (A_{T_i} - A_{S_i})\}\right|^2\right) \\ &= E\left(\sum_{i=1}^{n}\left|\left\{(A_{T_i}^{(n)} - A_{S_i}^{(n)}) - (A_{T_i} - A_{S_i})\right\}\right|^2\right) \\ &= E\left(\sum_{i=1}^{n}\int_{S_i}^{T_i}(u^{(n)} - u)_t^2 d\langle X\rangle_t\right) \\ &\le E\left(\int_0^1 (u^{(n)} - u)_t^2 d\langle X\rangle_t\right), \end{aligned}$$

hence showing that $A^{(n)}$ variationally converges to A. □

From now onwards a $\delta(\omega)$-fine belated partial stochastic division $D = \{((S_i, T_i], \xi_i)\}_{i=1}^n$ of $[0, 1]$ will be denoted by $D = \{((S, T], \xi)\}$, in which $(S, T]$ represents a typical stochastic interval in D, namely, $(S_i, T_i]$ and $((S, T], \xi)$ a typical stochastic interval-point pair, namely, $((S_i, T_i], \xi_i)$.

Theorem 4.12 (Variational Convergence Theorem). *Let u, $u^{(n)}$, $n = 1, 2, \ldots$, be adapted processes on the standard filtering space $(\Omega, \mathcal{F}, \{\mathcal{F}_t\}, P)$ such that $u^{(n)}$ converges to u everywhere in $[0, 1]$ except on a set of μ_X-measure zero. Suppose that each $u^{(n)}$ is WHVB integrable to a process $A^{(n)}$ on $[0, 1]$ with respect to an L^2-martingale X; and $A^{(n)}$ variationally converges to A. Then u is WHVB integrable to the process A on $[0, 1]$, and*

$$\lim_{n\to\infty} E\left(\left|(WHVB)\int_0^1 u_t^{(n)} dX_t\right|^2\right) = E\left(\left|(WHVB)\int_0^1 u_t dX_t\right|^2\right).$$

Proof. Let $(E(f^2))^{1/2}$ be denoted by $\|f\|$ and $X_T - X_S$ be denoted by $X(S, T)$. Given $\epsilon > 0$, by the variational convergence of $A^{(n)}$, we may assume that for each n, we have

$$\left\|(D)\sum\{(A_T^{(n)} - A_S^{(n)}) - (A_T - A_S)\}\right\| < \frac{\epsilon}{2^n} \tag{4.5}$$

for any finite collection of disjoint stochastic intervals $D = \{(S, T]\}$. Otherwise, in our proof, we may replace n by a subsequence $n_k, k = 1, 2, \ldots$, where for each n_k we have

$$\left\|(D)\sum\{(A_T^{(n_k)} - A_S^{(n_k)}) - (A_T - A_S)\}\right\| < \frac{\epsilon}{2^k}.$$

Furthermore, for each n, there exists a locally stopping process $\delta^n(\xi,\omega) > 0$ such that for any δ^n-fine belated partial stochastic division $D_n = \{((S,T],\xi)\}$, we have

$$\left\|(D_n)\sum\{u_\xi^{(n)}X(S,T) - A^{(n)}(S,T)\}\right\| < \frac{\epsilon}{2^n}.$$

By Lemma 4.7, we may assume that $u^{(n)}$ converges to u everywhere on $[0,1]\times\Omega$. By Lemma 4.8, there exists an adapted process $n(\xi,\omega)$ such that

$$\left|u^{n(\xi,\omega)}(\xi,\omega) - u(\xi,\omega)\right| < \epsilon$$

for each $(\xi,\omega) \in [0,1]\times\Omega$. Define $\delta(\xi,\omega) = \delta^{n(\xi,\omega)}(\xi,\omega)$. By Lemma 4.9, $\delta(\xi,\omega)$ is a locally stopping process. Let $D = \{((S,T],\xi)\}$ be any $\delta(\omega)$-fine belated partial stochastic division of $[0,1]\times\Omega$. Then

$$\begin{aligned}
&E\left(\left|(D)\sum\{u_\xi X(S,T) - A(S,T)\}\right|^2\right)\\
&\le 3E\left(\left|(D)\sum(u(\xi,\omega) - u^{n(\xi,\omega)}(\xi,\omega))(X(T(\omega),\omega) - X(S(\omega),\omega))\right|^2\right)\\
&\quad + 3E\Bigg(\Bigg|(D)\sum\{u^{n(\xi,\omega)}(\xi,\omega)(X(T(\omega),\omega) - X(S(\omega),\omega))\\
&\qquad\qquad - (A^{n(\xi,\omega)}(T(\omega),\omega) - A^{n(\xi,\omega)}(S(\omega),\omega)\}\Bigg|^2\Bigg)\\
&\quad + 3E\Bigg(\Bigg|(D)\sum\{(A^{n(\xi,\omega)}(T(\omega),\omega) - A^{n(\xi,\omega)}(S(\omega),\omega))\\
&\qquad\qquad - (A(T(\omega),\omega) - A(S(\omega),\omega))\}\Bigg|^2\Bigg)\\
&= 3(I_1 + I_2 + I_3).
\end{aligned}$$

Considering each of the above three parts separately, first we get

$$\begin{aligned}
I_1 \le \epsilon E\left|(D)\sum[X_T - X_S]\right|^2 &\le \epsilon E\left((D)\sum|X_T - X_S|^2\right)\\
&\le \epsilon\left(E\langle X\rangle_1 - E\langle X\rangle_0\right).
\end{aligned}$$

Note that $n(\xi,\omega)$ takes only positive integers. Assume that the range of $n(\xi,\omega)$ is $\{n_j : j = 1,2,\ldots\}$. Observe that $n(\xi,\omega_1)$ may not equal to $n(\xi,\omega_2)$ for $\omega_1 \neq \omega_2$. However, for fixed ξ, $G(\xi,n_j) = \{\omega : n(\xi,\omega) = n_j\} \in \mathcal{F}_\xi$, by Lemma 4.9. Note that $G(\xi,n_j)$ may not equal Ω. For each n_j, define $T_j(\omega) = T(\omega)$ and $S_j(\omega) = S(\omega)$ if $\omega \in G(\xi,n_j)$; otherwise

$T_j(\omega) = S_j(\omega) = \xi$. Then $D_j = \{((S_j, T_j], \xi)\}$ is a $\delta^{(n_j)}$-fine belated partial stochastic division of $[0, 1]$. Hence

$$I_2 = E\left(\left|\sum_{n_j}(D_j)\{u_\xi^{(n_j)}X(S_j, T_j) - A^{(n_j)}(S_j, T_j)\}\right|^2\right),$$

i.e.,

$$I_2^{\frac{1}{2}} \leq \sum_{n_j}\left\|(D_j)\sum\{u_\xi^{(n_j)}X(S_j, T_j) - A^{(n_j)}(S_j, T_j)\}\right\| < \sum_{n_j}\frac{\epsilon}{2^{n_j}} \leq \epsilon.$$

As in the case of I_2, we get

$$I_3^{\frac{1}{2}} < \sum_{n_j}\frac{\epsilon}{2^{n_j}} \leq \epsilon.$$

Hence u is WHVB integrable to the process A on $[0, 1]$. Furthermore, from Equation (4.5),

$$\lim_{n\to\infty}\left\|\{(A_1^{(n)} - A_0^{(n)}) - (A_1 - A_0)\}\right\| = 0.$$

Hence $\lim_{n\to\infty}\|(A_1^{(n)} - A_0^{(n)})\| = \|(A_1 - A_0)\|$. Thus,

$$\lim_{n\to\infty} E\left(\left|(WHVB)\int_0^1 u_t^{(n)}dX_t\right|^2\right) = E\left(\left|(WHVB)\int_0^1 u_t dX_t\right|^2\right).$$

□

We are now ready to prove the Equivalence Theorem of the classical Stochastic Integral and WHVB Integral.

Theorem 4.13 (Equivalence Theorem). *If u is classical stochastic integrable with respect to an L^2-martingale X on $[0, 1]$, then u is WHVB integrable on $[0, 1]$, and the integrals agree.*

Proof. By Lemma 4.6, the theorem holds true for an adapted simple process u. Now suppose $u \in \mathcal{L}_2^{\langle X\rangle}$. Then there exists a sequence of adapted simple processes $\{u^{(n)}\}$ in $\mathcal{L}_0$ such that $u^{(n)}$ converges to u on $[0, 1] \times \Omega$ except a set of μ_X-measure zero, and $\lim_{n\to\infty} E(\int_0^1 (u^{(n)} - u)_t^2 d\langle X\rangle_t) = 0$, see [Chung and Williams (1990), p. 37] and [Weizsacker and Winkler (1990), p. 120].

Let $A^{(n)}(S, T) = (I)\int_S^T u_t^{(n)}dX_t = (WHVB)\int_S^T u_t^{(n)}dX_t$ and $A(S, T) = (I)\int_S^T u_t dX_t$. Then by Example 4.4, $A^{(n)}$ variationally converges to A. Hence by Theorem 4.12, u is WHVB integrable to A on $[0, 1]$. □

Theorem 4.14 (Isometry). *Let* $u \in \mathcal{L}_2^{\langle X \rangle}$. *Then*

$$E\left(\left|(WHVB)\int_0^1 u_t dX_t\right|^2\right) = E\left(\int_0^1 u_t^2 d\langle X\rangle_t\right).$$

Proof. First, we shall show that the isometry equality holds for any adapted simple process $u \in \mathcal{L}_0$, i.e.,

$$u_t = \sum_{j=0}^{m-1} f_j 1_{(s_j, s_{j+1}]}(t)$$

where $0 = s_0 < s_1 < s_2 < \cdots < s_m = 1$, each f_j is $\mathcal{F}_{s_j}$-measurable and there exists $M > 0$ such that $|f_j(\omega)| \leq M$ for all $\omega \in \Omega$ and all j.

$$\begin{aligned}
E\left(\left|(WHVB)\int_0^1 u_t dX_t\right|^2\right) &= E\left(\sum_{j=0}^{m-1} f_j(X_{s_{j+1}} - X_{s_j})\right)^2 \\
&= E\left(\sum_{k=0}^{m-1} f_k^2 (X_{s_{k+1}} - X_{s_k})^2\right) \\
&\quad + E\left(\sum_{i \neq j} f_i f_j (X_{s_{i+1}} - X_{s_i})(X_{s_{j+1}} - X_{s_j})\right) \\
&= E\left(\sum_{k=0}^{m-1} f_k^2 (X_{s_{k+1}} - X_{s_k})^2\right) + 0 \\
&= E\left(\sum_{k=0}^{m-1} f_k^2 \left(\langle X\rangle_{s_{k+1}} - \langle X\rangle_{s_k}\right)\right) \\
&= E\left(\int_0^1 u_t^2 d\langle X\rangle_t\right).
\end{aligned}$$

Now suppose $u \in \mathcal{L}_2^{\langle X \rangle}$. Then there exists a sequence of adapter simple processes $\{u^{(n)}\}$ in $\mathcal{L}_0$ such that

$$\lim_{n\to\infty} E\left(\int_0^1 \left(u_t^{(n)}\right)^2 d\langle X\rangle_t\right) = E\left(\int_0^1 u_t^2 d\langle X\rangle_t\right).$$

On the other hand, the isometry property holds for $u^{(n)}$ for all n,

$$E\left((WHVB)\int_0^1 u_t^{(n)} dX_t\right)^2 = E\left(\int_0^1 \left(u_t^{(n)}\right)^2 d\langle X\rangle_t\right).$$

Thus

$$\lim_{n\to\infty} E\left((WHVB)\int_0^1 u_t^{(n)} dX_t\right)^2 = \lim_{n\to\infty} E\left(\int_0^1 \left(u_t^{(n)}\right)^2 d\langle X\rangle_t\right)$$
$$= E\left(\int_0^1 u_t^2 d\langle X\rangle_t\right).$$

By Theorems 4.12 and 4.13,

$$E\left(\left|(WHVB)\int_0^1 u_t dX_t\right|^2\right) = E\left((WHVB)\int_0^1 u_t^2 d\langle X\rangle_t\right).$$

□

4.3.2 *Local L^2-martingale as integrator*

Theorem 4.15. *Let X be a local L^2-martingale with a corresponding sequence of stopping times $(\tau_n), n = 1, 2, \ldots$, see Definition 1.11, Chapter 1, for the definition of a local martingale. Suppose u is WHVB integrable to a process A with respect to X. Then u is WHVB integrable to a process $A^{(n)}$ with respect to $X^{(n)}$ for each $n = 1, 2, \ldots$, where $X^{(n)} = \{X_{t\wedge\tau_n} : t \in [0,1]\}$ and $A^{(n)} = \{A_{t\wedge\tau_n} : t \in [0,1]\}$. Furthermore, for each $t \in [0,1]$, $\lim_{n\to\infty} A^{(n)}(t,\omega) = A(t,\omega)$ a.s.*

Proof. Let $\epsilon > 0$ and $\delta(\xi,\omega) > 0$ be given as in Definition 4.6 for the WHVB integral of u with respect to X. Let $\delta^{(n)}(\xi,\omega) > 0$ be defined as follows: $\delta^{(n)}(\xi,\omega) = \tau_n(\omega) - \xi$ if $\xi < \tau_n(\omega) \le \xi + \delta(\xi,\omega)$; $\delta^{(n)}(\xi,\omega) = \delta(\xi,\omega)$ otherwise. Then $\delta^{(n)}(\xi,\omega)$ is a locally stopping process in view of the fact that $(\tau_n(\omega)\wedge\xi)\wedge(\xi+\delta(\xi,\omega))$ is a stopping time and $\{\omega : \tau_n(\omega) < \xi\} \in \mathcal{F}_t$ if $\xi < t$. Observe that if $((S,T],\xi)$ is a $\delta^{(n)}$-fine belated stochastic interval-point pair, then for the case $\xi < \tau_n(\omega) \le \xi + \delta(\xi,\omega)$ or $\xi + \delta(\xi,\omega) < \tau_n(\omega)$, we have

$$X^{(n)}(T(\omega),\omega) - X^{(n)}(S(\omega),\omega) = X(T(\omega)\wedge\tau_n(\omega),\omega) - X(S(\omega)\wedge\tau_n(\omega),\omega)$$
$$= X(T(\omega),\omega) - X(S(\omega),\omega);$$

for the case $\tau_n(\omega) \le \xi$, we have

$$X^{(n)}(T(\omega),\omega) - X^{(n)}(S(\omega),\omega) = X(\tau_n(\omega),\omega) - X(\tau_n(\omega),\omega) = 0.$$

Similar results hold for $A^{(n)}$. Furthermore, $((S,T],\xi)$ is also δ-fine belated. Hence u is WHVB integrable to a process $A^{(n)}$ with respect to $X^{(n)}$. It is clear that for each $t \in [0,1]$, $\lim_{n\to\infty} A^{(n)}(t,\omega) = A(t,\omega)$ a.s. □

From Theorems 4.15 and 4.13, we have the following:

Theorem 4.16. *If u is classical stochastic integrable with respect to a local L^2-martingale X on $[0,1]$, then it is WHVB integrable on $[0,1]$ and the two integrals agree.*

4.3.3 *Semimartingale as integrator*

Remark 4.6. We shall point out that the Definition 4.6 can also be applied to an integrator $X = \{X_t(\omega) : t \in [0,1]\}$, which is adapted, cadlag and has paths of finite variation on $[0,1]$. For each $\omega \in \Omega$, let $V_t(\omega)$ be the total variation of $X(\cdot,\omega)$ over $[0,t]$. First we assume that $\{V_t\}$ is an L^2-process (which can be omitted as pointed out later). Let $B_2(X)$ be the space of L^2-processes $u = \{u_t\}$ which is a predictable process on $(\Omega, \mathcal{F}, \{\mathcal{F}_t\}, P)$ such that for a.s. ω, $u_t(\omega)$ is Lebesgue–Stieltjes integrable with respect to $X_t(\omega)$ on $[0,1]$ and $E\left|\int_0^1 u_t dX_t\right|^2 < \infty$ (which can be weakened below). Then Lemmas 4.5–4.9 and Theorems 4.12–4.13 also hold true for $B_2(X)$. However, in the proof of Theorem 4.12 (Variational Convergence Theorem),

$$I_1 \leq \epsilon E\left(\left|(D)\sum(X_T - X_S)\right|^2\right) \leq \epsilon E(V_1^2).$$

Now observe that as in local L^2-martingales, by introducing a suitable sequence of stopping times, we can omit the condition that $\{V_t\}$ is an L^2-process and weaken the condition $E\left|\int_0^1 u_t dX_t\right|^2 < \infty$ by replacing it by $\int_0^1 u_t dX_t$ exists a.s. In other words, we too can consider these two conditions locally with respect to a suitable sequence of stopping times.

So the WHVB integral and the Lebesgue–Stieltjes integral agree. Recall that a semimartingale S can be written as $S = L + B$, where L is a local martingale and B is a process which has paths of finite variation. Hence the WHVB integral and the classical stochastic integral with respect to a semimartingale agree.

4.3.4 *Classical Itô's formula*

The classical Itô's formula

$$\int_a^b F'(X_t)dX_t = F(X_b) - F(X_a) - \frac{1}{2}\int_a^b F''(X_t)d\langle X\rangle_t$$

is well-known, where X is a continuous L^2-martingale, the integral on the left is the stochastic integral and that on the right is the Lebesgue–Stieltjes integral, see, e.g. [Klebaner (2012), pp. 222, 236]. This formula is useful for evaluating certain stochastic integrals since it links stochastic integral to the classical Stieltjes integral, which is usually easy to evaluate. In this section, we shall use the variational approach to offer a new proof of a version of the Itô's formula for L^2-martingale, Theorem 4.17.

We need the following lemmas.

Lemma 4.10. *Let $\varphi = \{\varphi(t,\omega) : t \in [0,1]\}$ be a cadlag process and which is adapted to $(\Omega, \mathcal{F}, \{\mathcal{F}_t\}, P)$. Let X be an adapted process with finite quadratic variation. Then given any $\epsilon > 0$, there exists a stopping process $\delta(\xi,\omega) > 0$ such that for any $\delta(\omega)$-fine belated partial stochastic division $D = \{(U,V),\xi\}$ of $[0,1] \times \Omega$, we have*

$$E\left|(D)\sum\{[\varphi_\Theta - \varphi_\xi][X_V - X_U]\}\right|^2 < \epsilon$$

where $\Theta \in (U,V]$, that is, $\Theta(\omega) \in (U(\omega), V(\omega))$ for each $\omega \in \Omega$.

Proof. Given $\epsilon > 0$, by Lemma 4.3, we can choose a stopping process $\delta(\xi,\omega) > 0$ such that for a fixed ξ and for any $u(\omega) \in [0,1]$ with $0 < u(\omega) - \xi < \delta(\xi,\omega)$ we have

$$|\varphi(u(\omega),\omega) - \varphi(\xi,\omega)| < \frac{\sqrt{\epsilon}}{\sqrt{V(X)}}$$

where $V(X)$ denotes the quadratic variation of X (see Definition 4.9). For any $\delta(\omega)$-fine belated partial stochastic division $D = \{(U,V],\xi\}$, and corresponding partial division $\{(U,V],\Theta\}$ with $\Theta \in (U,V)$, i.e., $\Theta(\omega) \in (U(\omega),V(\omega)) \subset (\xi, \xi + \delta(\xi,\omega))$, we have

$$\begin{aligned} E\left|(D)\sum[\varphi_\Theta - \varphi_\xi][X_V - X_U]\right|^2 &< \frac{\epsilon}{V(X)} E\left|\sum[X_V - X_U]\right|^2 \\ &\le \frac{\epsilon}{V(X)} \cdot V(X) = \epsilon \end{aligned}$$

thereby completing the proof. □

Lemma 4.11. *Let G be a bounded process on $[0,1]$ and X be a cadlag L^2-martingale. Given any $\epsilon > 0$ there exists a stopping process $\delta(\xi,\omega) > 0$ such that for any $\delta(\omega)$-fine belated partial stochastic division $D = \{((U,V],\xi)\}$ with ξ replaced by any $\Theta \in (U,V]$, i.e., $\Theta(\omega) \in (U(\omega),V(\omega))$, we have*

$$E\left|(D)\sum G_\Theta\left\{[X_V - X_U]^2 - (\langle X\rangle_V - \langle X\rangle_U)\right\}\right|^2 < \epsilon.$$

Proof. Given $\epsilon > 0$ choose a stopping process $\delta(\xi,\omega) > 0$ such that for any u satisfying $0 < u - \xi < \delta(\xi,\omega)$, we have

$$|X(u,\omega) - X(\xi,\omega)| < \frac{1}{2}\sqrt{\epsilon}$$

and

$$|\langle X\rangle(v,\omega) - \langle X\rangle(\xi,\omega)| < \frac{1}{2}\sqrt{\epsilon}.$$

Let $M > 0$ be such that $|G(u,\omega)| \leq \sqrt{M}$ for all $(u,\omega) \in [0,1] \times \Omega$. Let $D = \{(U,V),\xi\}$ be any $\delta(\omega)$-fine belated partial stochastic division. Then

$$\begin{aligned}I &= E\left|(D)\sum G_{\Theta}\left\{[X_V - X_U]^2 - (\langle X\rangle_V - \langle X\rangle_U)\right\}\right|^2\\ &\leq ME\left|(D)\sum \left\{(X_V - X_U)^2 - (\langle X\rangle_V - \langle X\rangle_U)\right\}\right|^2\\ &= M\sum E\left\{[X_V - X_U]^2 - (\langle X\rangle_V - \langle X\rangle_U)\right\}^2\end{aligned}$$

since the process $[X_V - X_U]^2 - (\langle X\rangle_V - \langle X\rangle_U)$ has orthogonal increment, as

$$\begin{aligned}&E\{\left([X_{V_i} - X_{U_i}]^2 - (\langle X\rangle_{V_i} - \langle X\rangle_{U_i})\right)\left([X_{V_j} - X_{U_j}]^2 - (\langle X\rangle_{V_j} - \langle X\rangle_{U_j})\right)\}\\ &\quad = E\left\{E\left\{\left([X_{V_i} - X_{U_i}]^2 - (\langle X\rangle_{V_i} - \langle X\rangle_{U_i})\right)\right.\right.\\ &\qquad \left.\left.\left([X_{V_j} - X_{U_j}]^2 - (\langle X\rangle_{V_j} - \langle X\rangle_{U_j})\right)|\mathcal{F}_{U_i}\right\}\right\}\\ &\quad = E\left\{\left([X_{V_i} - X_{U_i}]^2 - (\langle X\rangle_{V_i} - \langle X\rangle_{U_i})\right)\right.\\ &\qquad \left.E\left\{\left([X_{V_j} - X_{U_j}]^2 - (\langle X\rangle_{V_j} - \langle X\rangle_{U_j})\right)|\mathcal{F}_{U_i}\right\}\right\}\\ &\quad = 0.\end{aligned}$$

So

$$\begin{aligned}I &\leq M\Big\{\sum E[X_V - X_U]^4 + \sum(E\langle X\rangle_V - E\langle X\rangle_u)^2\\ &\qquad - 2\sum E([X_V - X_U]^2(\langle X\rangle_V - \langle X\rangle_U))\Big\}^2\\ &\leq M\Big\{\epsilon\sum E(X_V - X_U)^2 + \epsilon\sum(E\langle X\rangle_V - E\langle X\rangle_U)\\ &\qquad + 2\epsilon\sum(E\langle X\rangle_V - E\langle X\rangle_U)\Big\}\\ &< 4M(E\langle X\rangle_1 - E\langle X\rangle_0)\epsilon,\end{aligned}$$

thereby completing the proof. □

Lemma 4.12. *Let G be an adapted cadlag process on $[0,1]$. Then, for any given $\epsilon > 0$ there exists positive stopping process $\delta(\xi,\omega) > 0$ such that* *for any* $\delta(\omega)$*-fine belated partial stochastic division* $D = \{((U,V],\xi)\}$ *of* $[0,1]\times\Omega$ *we have*

$$E\left|(D)\sum\left\{G_\Theta(\langle X\rangle_V - \langle X\rangle_U) - \int_U^V G(t,\omega)d\langle X\rangle_t\right\}\right|^2 < \epsilon.$$

Proof. Let $\epsilon > 0$ be given. Choose a positive stopping process $\delta(\xi,\omega) > 0$ such that

$$|\langle X\rangle(V(\omega),\omega) - \langle X\rangle(U(\omega),\omega)| < \epsilon$$

and

$$|G(V(\omega),\omega) - G(U(\omega),\omega)| < \sqrt{\epsilon},$$

where $(U,V]$ is belated $\delta(\omega)$-fine, i.e., $(U(\omega),V(\omega)] \subseteq (\xi,\xi+\delta(\xi,\omega))$. Choose a $\delta(\omega)$-fine belated partial stochastic division $D = \{((U,V),\xi)\}$. Let $\Theta(\omega) \in (U(\omega),V(\omega)]$. Define

$$G_D(t,\omega) = G(\Theta(\omega),\omega) \text{ if } t \in (U(\omega),V(\omega)).$$

Then

$$\begin{aligned}
J &= E\left|(D)\sum\left\{G_\Theta(\langle X\rangle_V - \langle X\rangle_U) - \int_U^V G(t,\omega)d\langle X\rangle_t\right\}\right|^2\\
&= E\left|(D)\sum\int_U^V [G_D(t,\omega) - G(t,\omega)]d\langle X\rangle_t\right|^2\\
&< \epsilon E\left|\sum(\langle X\rangle_V - \langle X\rangle_U)\right|^2\\
&\leq \epsilon\left|E(\langle X\rangle_1 - \langle X\rangle_0)\right|^2
\end{aligned}$$

thereby completing our proof. □

Theorem 4.17 (Itô's Formula). *Let* X *be a cadlag* L^2*-martingale and* $F:\mathbb{R}\to\mathbb{R}$ *be continuous and twice differentiable such that*

(i) $F'\circ X$ *is WHVB integrable with respect to* X*; and*
(ii) $F''\circ X$ *is bounded.*

Then

$$(WHVB)\int_0^1 F'(X_t)dX_t = F(X_1) - F(X_0) - \frac{1}{2}\int_0^1 F''(X_t)d\langle X\rangle_t.$$

Proof. Since F is twice differentiable for each $\omega \in \Omega$ by Taylor's Theorem we have

$$F(X_V) = F(X_U) + [X_V - X_U]F'(X_U) + \frac{1}{2}[X_V - X_U]^2 F''(X_\Theta)$$

for some $\Theta \in (U, V]$.

For any stochastic interval $(U, V]$ of $[0, 1] \times \Omega$, the process $G(U, V]$ is defined by

$$G(U, V] = F(X_V) - F(X_U) - \frac{1}{2}\int_U^V F''(X_t)d\langle X\rangle_t,$$

where $\int_U^V F''(X_t)d\langle X\rangle_t$ is Lebesgue–Steltjes which exists. Hence

$$G(U, V] = [X_V - X_U]F'(X_U) + \frac{1}{2}(X_V - X_U)^2 F''(X_\Theta) - \frac{1}{2}\int_U^V F''(X_t)d\langle X_t\rangle.$$

Choose a stopping process $\delta(\xi, \omega)$ to be determined later. The existence of such a stopping process is guaranteed by Lemmas 4.11 and 4.12. Consider any $\delta(\omega)$-fine belated partial stochastic division $D = \{((U, V], \xi)\}$ of $[0, 1]$.

We have

$$\begin{aligned}
K &= E\left|(D)\sum\{F'(X_\xi)[X_V - X_U] - G(U, V]\}\right|^2 \\
&= E\left|(D)\sum\left\{F'(X_\xi)[X_V - X_U] - F'(X_U)[X_V - X_U]\right.\right. \\
&\qquad \left.\left. - \frac{1}{2}[X_V - X_U]^2 F''(X_\Theta) + \frac{1}{2}\int_U^V F''(X_t)d\langle X\rangle_t\right\}\right|^2 \\
&\le 2E\left|(D)\sum[F'(X_\xi) - F'(X_U)[X_V - X_U]\right|^2 \\
&\quad + 2\left\{2E\left|(D)\sum\{F''(X_\Theta)[X_V - X_U]^2 - F''(X_\Theta)(\langle X\rangle_V - \langle X\rangle_U)\}\right|^2\right. \\
&\qquad \left. + 2E\left|(D)\sum\{F''(X_\Theta)(\langle X\rangle_V - \langle X\rangle_U) - \int_U^V F''(X_t)d\langle X\rangle_t\}\right|^2\right\} \\
&< 2\epsilon + 4\epsilon + 4\epsilon \\
&= 10\epsilon
\end{aligned}$$

if we choose $\delta(\xi, \omega) > 0$ to satisfy Lemma 4.10 with φ replaced by $F' \circ X$, and G in Lemma 4.11 replaced by $F'' \circ X$ and G in Lemma 4.12 replaced by $F'' \circ X$. □

We can generalize the above result to the case when X is a local L^2-martingale.

Theorem 4.18 (Itô's Formula for Local Martingale). *Let X be a cadlag local L^2-martingale and $F : \mathbb{R} \to \mathbb{R}$ be continuous and twice differentiable such that*

(i) *$F' \circ X$ is WHVB integrable with respect to X; and*
(ii) *$F'' \circ X$ is bounded.*

Then

$$(WHVB)\int_0^1 F'(X_t)dX_t = F(X_1) - F(X_0) - \frac{1}{2}\int_0^1 F''(X_t)d\langle X\rangle_t.$$

Proof. Now $F' \circ X$ is WHVB integrable with respect to X implies that $F' \circ X$ is WHVB integrable with respect to $X^{(n)}$ for each $n = 1, 2, \ldots$, where each $X^{(n)}$ is defined in Definition 1.11 and is a L^2-martingale. Apply Theorem 4.17 to each n, we have

$$(WHVB)\int_0^1 F'(X_t)dX_t^{S_n\wedge} = F(X_1) - F(X_0) - \frac{1}{2}\int_0^1 F''(X_t)d\langle X^{S_n\wedge}\rangle_t.$$

Again by Definition 1.11,

$$\begin{aligned}(WHVB)\int_0^1 F'(X_t)dX_t &= \lim_{n\to\infty}\int_0^1 F'(X_t)dX_t^{S_n\wedge} \\ &= \lim_{n\to\infty} F(X_1) - F(X_0) - \frac{1}{2}\int_0^1 F''(X_t)d\langle X^{S_n\wedge}\rangle_t \\ &= F(X_1) - F(X_0) - \frac{1}{2}\int_0^1 F''(X_t)d\langle X\rangle_t\end{aligned}$$

thereby completing our proof. □

4.4 Notes and Remarks

Results using the variational approach and their proofs in this chapter can be found in [Chew, Tay and Toh (2001–02); Toh and Chew (1999, 2002, 2005, 2012)].

We have not attempted to define differentials for L^2-martingales and semimartingales in Chapter 4. Using ideas in Chapter 3, Section 3.2, we may able to define differential for L^2-martingales and semimartingales, and as a consequence, we may give another proof of Itô's formula for local L^2-martingales.

Variational integrals for deterministic functions can be found in [Henstock (1988), p. 54; Henstock (1991), p. 87; Lee (1989), p. 51; Muldowney (2012), p. 146].

Chapter 5

The Multiple Itô–Wiener Integral

The theory of Multiple Stochastic Integral was first introduced by N. Wiener in 1938, who used the term *polynomial chaos.* This study was followed up by K. Itô in 1951, see [Itô (1951)]. Itô's approach of studying the Multiple Stochastic Integral was a generalization of his approach to the one-dimensional Wiener integral, in which case the integrator is a canonical Brownian motion and the integrand is a real-valued *deterministic* function. His one-dimensional integral is what we call the *Itô–Wiener Integral* while his generalization to m-dimension is the *Multiple Itô–Wiener Integral.*

In this chapter, we attempt to use McShane's approach to define a Multiple Stochastic Integral in m-dimensional Euclidean space $[0,1]^m$. We show that the Itô–McShane integral defined here is equivalent to the classical Multiple Itô–Wiener integral where the Multiple Itô–McShane integral does not take into the account the values of the integrand on the diagonal.

We shall study the integral on the diagonal, called the Multiple Wiener–McShane integral, and derive Fubini's Theorem and the classical Hu–Meyer Theorem using McShane setting in Chapter 6.

We shall first recall McShane's full division on $[0,1]$. McShane's approach in defining non-stochastic integral on $[0,1]$ is well-known. It uses a full division with non-uniform mesh induced by $\delta(\xi) > 0$. The full δ-fine McShane's division $D\{([u,v],\xi)\}$ consists of interval-point pair for which $[u,v] \subset (\xi - \delta(\xi), \xi + \delta(\xi))$ and where the associated point ξ may or may not lie in $[u,v]$. The difference between McShane's division and Henstock's division is that the *associated point* ξ in the latter is contained within the interval but the associated point ξ in former need not be in the interval.

It was proved that McShane's integral is equivalent to the classical Lebesgue integral for the non-stochastic case. To be consistent with the conventions for the subsequent sections, we shall use the corresponding left-open division $D = \{((u,v],\xi)\}$ in place of closed division $D = \{([u,v],\xi)\}$, so that our division covers $(0,1]$ instead of $[0,1]$.

As mentioned in Chapter 1, the integral using this approach is not equivalent to the classical stochastic integral even when the integrator is a canonical Brownian motion.

5.1 Definition of the Integral

In this section, we shall generalize the theory of stochastic integration to m-dimensional Euclidean space for the deterministic real-valued function $f : [0,1]^m \to \mathbb{R}$. The integrator is the Brownian motion in one-dimension, $B = \{B_t(\omega) : t \in [0,1]\}$ as defined in Chapter 1. Let $T = [0,1]$ and $T^m = [0,1]^m$. We shall denote the Lebesgue measure on T by λ. Let λ^m be the corresponding Lebesgue measure on T^m. For any interval I of T, we may use $|I|$ or $\lambda(I)$ to denote the length of I.

It could be seen that T^m consists of two parts, namely, the diagonal part of T^m

$$\mathcal{D} = \{(x_1, x_2, \ldots, x_m) \in T^m : x_i = x_j \text{ for some } i \neq j\},$$

and

$$\mathcal{D}^c = \{(x_1, x_2, \ldots, x_m) \in T^m : x_i \neq x_j \text{ for any } i \neq j\},$$

which is the non-diagonal part of T^m.

The non-diagonal set can be decomposed to $m!$ open connected sets in T^m. For each $\pi \in S_m$, the group of all permutations of m objects, we define

$$G_\pi = \{(x_1, x_2, \ldots, x_m) \in T^m : x_{\pi(1)} < x_{\pi(2)} < \cdots < x_{\pi(m)}\},$$

and there are $m!$ such sets. Each of these sets is said to be *contiguous* to the diagonal $\mathcal{D}$.

The motivation to divide T^m into two parts, namely, the diagonal and non-diagonal parts, occur from the classical integration theory. It is well-known in classical integration theory that if a set A has Lebesgue measure zero, then the Lebesgue integral of f over A is zero. In Chapter 2, we also have shown that in the one-dimensional space, if A is a set of Lebesgue measure zero on $[0,1]$, then $\int_A f dB_t = 0$. A natural question to ask is: can this result be generalized to m-dimensional space? It turns out that if A has null intersection with the diagonal, the answer is positive; however, if A has non-empty intersection with the diagonal, the answer is otherwise.

Let f be a real-valued function on T^m. Define a function $f_0(x) = f(x)$ if $x \in \mathcal{D}^c$; otherwise, $f_0(x) = 0$. Then f_0 is called the non-diagonal part of f.

Definition 5.1. Let δ be a positive function defined on T^m, $\xi = (\xi_1, \xi_2, \ldots, \xi_m) \in T^m$ and $I = \prod_{i=1}^m I_i$ be an interval of T^m. An *interval-point pair* (I, ξ) is said to be McShane's δ-fine if $I_k \subset (\xi_k - \delta(\xi), \xi_k + \delta(\xi))$ for each $k = 1, 2, \ldots, m$.

A finite collection D of McShane's δ-fine interval-point pairs $\{(I^{(i)}, \xi^{(i)})\}_{i=1}^n$ is said to be a McShane's δ-fine division of T^m if

(i) $\{I^{(i)}\}$, $i = 1, 2, \ldots, n$, and are disjoint intervals of T^m; and
(ii) $\bigcup_{i=1}^{n} I^{(i)} = (0, 1]^m$.

Remark 5.1.

(i) The intervals I that we use are left-open in each of its components, that is, $I = \prod_{i=1}^m I_i$, where I_i is of the form $(a_i, b_i]$.
(ii) Note that ξ_k may or may not be in I_k for each $k = 1, 2, \ldots, m$.
(iii) We remark that for any given positive function δ on T^m, a δ-fine division of T^m exists.
(iv) In this chapter, we always use McShane's δ-fine divisions. For brevity, a McShane's δ-fine division is often called a δ-fine division in this chapter.

Let f be a real-valued function on T^m. If $D = \{(I^{(i)}, x^{(i)})\}$ is a δ-fine division of T^m, then we let $S(f, \delta, D)$ denote the Riemann sum

$$S(f, \delta, D) = \sum f(x^{(i)}) B(I^{(i)}),$$

where

$$B(I^{(i)}) = \prod_{j=1}^{m} B(I_j^{(i)})$$

if $I^{(i)} = \prod_{j=1}^m I_j^{(i)}$, each $I_j^{(i)}$ is a left-open interval of T and $B(I_j) = B_{b_j} - B_{a_j}$, where $I_j = (a_j, b_j]$.

Definition 5.2. A function $f : T^m \to \mathbb{R}$ is said to be Multiple Itô–McShane integrable to $IM(f) \in L^2(\Omega)$ if for every $\epsilon > 0$, there exists a positive function δ such that

$$E\left(|S(f_0, \delta, D) - IM(f)|^2\right) < \epsilon$$

whenever $D = \{(I^{(i)}, x^{(i)})\}_{i=1}^n$ is a δ-fine division of T^m.

A function $f : T^m \to \mathbb{R}$ is said to be Multiple Itô–McShane integrable on a subinterval I of T^m if $f 1_I$ is Multiple Itô–McShane integrable on T^m.

Note that in the Riemann sum $S(f_0, \delta, D)$, the integrand is the function f_0 instead of the function f, i.e., the values of f vanishes on the diagonal.

Lemma 5.1. *Let δ be a positive function on T^m and $\{D_k\}$ be a finite family of δ-fine divisions of T^m. Then there exists a partition $\{A_1, A_2, \ldots, A_q\}$ of $[0,1]$ and a finite family of δ-fine divisions of T^m denoted by $\{D'_k\}$ such that each interval of any D'_k is of the form $A_{l_1} \times A_{l_2} \times \cdots \times A_{l_m}$ and each D'_k is a refinement of D_k. Furthermore,*

$$S(f_0, \delta, D_k) = S(f_0, \delta, D'_k)$$

for all k.

Proof. It follows from the following fact: If (I, ξ) is δ-fine in T^m, and if $I = J \cup K$, where J and K are two disjoint left-open subintervals, then (J, ξ) and (K, ξ) are δ-fine. Furthermore,

$$f(\xi)B(I) = f(\xi)B(J) + f(\xi)B(K),$$

and by taking $\{A_1, A_2, \ldots, A_q\}$ to be all the intervals formed by taking all the division points of $\{D_k\}$, the proof is easily completed. □

We remark that Lemma 5.1 holds since δ-fine divisions used are McShane's δ-fine divisions: the tag may not be in the interval and more than one interval may have the same tag.

Lemma 5.1 also holds true for just one McShane's δ-fine division. We have the following definition.

Definition 5.3. Let $T^m = [0,1]^m$ and $D = \{(I, \xi)\}$ be a δ-fine division of T^m. Suppose there exists a partition $\{A_1, A_2, \ldots, A_q\}$ of $[0,1]$ and every interval I in D is of the form $A_{l_1} \times A_{l_2} \times \cdots \times A_{l_m}$. Then D is called a standard δ-fine division.

In view of Lemma 5.1, we shall assume that all finite family of δ-fine divisions of T^m that we consider in Definition 5.2 are standard divisions.

Theorem 5.1. *The Multiple Itô–McShane integral, if it exists, is unique.*

Proof. The proof is standard in the theory of Riemann integration, hence omitted. □

In view of Theorem 5.1, we may let $IM(f)$ denote the Multiple Itô–McShane integral of f for our subsequent sections.

Theorem 5.2 (Cauchy's Criterion). *A function f is Multiple Itô–McShane integrable on T^m if and only if given $\epsilon > 0$, there exists a positive function δ on T^m such that*

$$E\left(|S(f_0, \delta, D_1) - S(f_0, \delta, D_2)|^2\right) < \epsilon$$

whenever D_1 and D_2 are standard δ-fine divisions of T^m.

Proof. That the condition is necessary follows directly from the triangle inequality. We just need to prove the sufficiency. For each positive integer k, there is a positive function δ_k on $[a, b]$ such that the above inequality holds with $\epsilon = 1/k$. We may assume that $\delta_k(\xi) > \delta_{k+1}(\xi)$ for each k and every $\xi \in T^m$. For each k, let D_k be a δ_k-fine standard division of T^m. If $q > p$, then D_q and D_p are δ_p-fine. Then

$$E\left(|S(f_0, D_p, \delta_p) - S(f_0, D_q, \delta_q)|^2\right) < \frac{1}{p}.$$

Therefore $\{S(f_0, D_k, \delta_k)\}$ is a Cauchy sequence in $L^2(\Omega)$. Let A be the limit of $S(f_0, D_k, \delta_k)$ under L^2-norm. Then

$$E\left(|S(f_0, D_k, \delta_k) - A|^2\right) < \frac{1}{k}$$

for every k.

Let $\epsilon > 0$ and choose a positive integer N such that $\frac{1}{N} < \frac{\epsilon}{4}$. Suppose D is a δ_N-fine division of T^m. Then

$$\begin{aligned} E\left(|S(f_0, D, \delta_N) - A|^2\right) &\leq 2E\left(|S(f_0, D, \delta_N) - S(f_0, D_N, \delta_N)|^2\right) \\ &\quad + 2E\left(|S(f_0, D_N, \delta_N) - A|^2\right) \\ &\leq \frac{2}{N} + \frac{2}{N} \\ &< \epsilon. \end{aligned}$$

Thus f is Multiple Itô–McShane integrable on T^m. □

Theorem 5.3. *A function f is Multiple Itô–McShane integrable on T^m if and only if there exist positive functions δ_n on T^m, $n = 1, 2, \ldots$, with $\delta_{n+1} < \delta_n$ for all $n = 1, 2, \ldots$ and $\lim_{n\to\infty} \delta_n = 0$ such that $IM(f)$ is the limit of $S(f_0, \delta_n, D_n)$ under L_2-norm.*

Proof. Suppose $f : T^m \to \mathbb{R}$ is Multiple Itô–McShane integrable on T^m to $A \in L^2(\Omega)$. For each $\epsilon > 0$, there exists a positive function δ on T^m such that whenever D is a McShane δ-fine standard division of T^m, we have

$$E\left(|S(f_0, D, \delta) - A|^2\right) < \epsilon.$$

Consider $\epsilon = 1/n$ for $n = 1, 2, \ldots$. Let δ_n be the corresponding positive function. We may assume $\delta_{n+1}(\xi) < \delta_n(\xi)$ for all n and $\xi \in T^m$. Let D_n be any δ_n-fine standard division of $[a, b]$ for each n. Then we have

$$E\left(|S(f_0, D_n, \delta_n) - A|^2\right) < \frac{1}{n},$$

hence

$$\lim_{n\to\infty} E\left(|S(f_0, D_n) - A|^2\right) = 0.$$

Conversely, if there exists $A \in L^2(\Omega)$ and a decreasing sequence $\{\delta_n(\xi)\}$ of positive functions defined on T^m such that for any δ_n-fine division D_n of T^m, we have

$$\lim_{n\to\infty} E\left(|S(f_0, D_n, \delta_n) - A|^2\right) = 0.$$

Suppose f is not Multiple Itô–McShane integrable on T^m. Then there exists $\epsilon > 0$ such that for every positive function δ on T^m, there exists a McShane δ-fine standard division D of T^m with

$$E\left|S(f_0, D, \delta) - A\right|^2 \geq \epsilon.$$

Hence for each δ_n given above, there exists a δ_n-fine standard division D_n of T^m with

$$E\left|S(f_0, D_n, \delta_n) - A\right|^2 \geq \epsilon,$$

hence it leads to a contradiction. Therefore f is Multiple Itô–McShane integrable on T^m. □

5.2 Properties of the Integral

In this section, we shall state and prove the basic results of the Multiple Itô–McShane stochastic integral.

Theorem 5.4. *Let f_1 and f_2 be Multiple Itô–McShane integrable on T^m, and α be a real constant. Then*

(i) *$f_1 + f_2$ is Multiple Itô–McShane integrable on T^m and*

$$IM(f_1 + f_2) = IM(f_1) + IM(f_2);$$

(ii) *αf_1 is Multiple Itô–McShane integrable on T^m and*

$$IM(\alpha f_1) = \alpha IM(f_1).$$

(iii) *f_1 is Multiple Itô–McShane integrable on any subinterval of T^m.*

The proofs are standard in the theory of Henstock integration, hence omitted.

Definition 5.4.

(i) Let $f : T^m \to \mathbb{R}$ be a real-valued function. For each $\pi \in S_m$, where S_m is the set of all permutations on m objects, let f_π denote the *permuted function* of f under π, which is the function

$$f_\pi(t_1, t_2, \ldots, t_m) = f(t_{\pi(1)}, t_{\pi(2)}, \ldots, t_{\pi(m)})$$

for all $(t_1, t_2, \ldots, t_m) \in T^m$.

(ii) Let $f : T^m \to \mathbb{R}$ be a real-valued function. The *symmetrization* of f, denoted by $\tilde{f}$, is the function $\tilde{f} : T^m \to \mathbb{R}$ defined by

$$\tilde{f}(t_1, t_2, \ldots, t_m) = \frac{1}{m!} \sum_{\pi \in S_m} f_\pi(t_1, t_2, \ldots, t_m)$$

where the summation is done over all $\pi \in S_m$.

(iii) Let $f : T^m \to \mathbb{R}$ be a real-valued function. The function f is said to be symmetric if $f(t_1, t_2, \ldots, t_m) = f(t_{\pi(1)}, t_{\pi(2)}, \ldots, t_{\pi(m)})$ for any $\pi \in S_m$.

Theorem 5.5. *Let f be Multiple Itô–McShane integrable with value $IM(f)$ on T^m and let $\tilde{f}$ denote the symmetrization of the function f. Then $\tilde{f}$ is also Multiple Itô–McShane integrable on T^m and*

$$IM(f) = IM(\tilde{f}).$$

Proof. Given $\epsilon > 0$, there exists a positive function δ such that

$$E\left(\left|(D)\sum_{i=1}^{n} f_0(\xi^i) \prod_{r=1}^{m} B(I_{i_r}) - IM(f)\right|^2\right) < \frac{\epsilon}{(m!)^2}$$

for any δ-fine standard division $D = \{(\prod_{r=1}^m I_{i_r}, \xi^i)\}_{i=1}^n$ of T^m. For each $\pi \in S_m$, let f_{0_π} denote the *permuted function of f_0 under* π, that is,

$$f_{0_\pi}(t_1, t_2, \cdots, t_m) = f_0(t_{\pi(1)}, t_{\pi(2)}, \ldots, t_{\pi(m)})$$

for any $(t_1, t_2, \ldots, t_m) \in T^m$. Let δ_π be the permuted function of δ, so that considering any δ_π-fine standard division $D_\pi = \{(\prod_{r=1}^m I_{i_{\pi(r)}}, \xi_\pi^i)\}$ of T^m, we have

$$E\left(\left|(D_\pi)\sum_{i=1}^{n} f_{0_\pi}(\xi_\pi^i) \prod_{r=1}^{m} B(I_{i_{\pi(r)}}) - IM(f)\right|^2\right) < \frac{\epsilon}{(m!)^2}.$$

Note that $\prod_{r=1}^{m} B(I_{i_r}) = \prod_{r=1}^{m} B(I_{i_{\pi(r)}})$. Observe that $IM(f) = IM(f_\pi)$ for each π. Let

$$\delta(\xi) = \min_{\pi \in S_m} \delta_\pi(\xi) \text{ for all } \xi \in T^m.$$

For any δ-fine standard division $D = \{(\prod_{r=1}^{m} I_{i_r}, \xi^i)\}_{i=1}^{n}$ of T^m, we have

$$\begin{aligned}
E&\left(\left|(D)\sum_{i=1}^{n} \tilde{f}_0(\xi^i) \prod_{r=1}^{m} B(I_{i_r}) - IM(f)\right|^2\right) \\
&= E\left(\left|(D)\sum_{i=1}^{n}\left(\frac{1}{m!}\sum_{\pi \in S_m} f_{0_\pi}(\xi^i)\right)\prod_{r=1}^{m} B(I_{i_r}) - IM(f)\right|^2\right) \\
&= \frac{1}{(m!)^2} E\left(\left|(D)\sum_{\pi \in S_m}\left\{\sum_{i=1}^{n} f_{0_\pi}(\xi^i)\prod_{r=1}^{m} B(I_{i_r}) - IM(f)\right\}\right|^2\right) \\
&\leq \frac{1}{(m!)^2}(m!)\sum_{\pi \in S_m} E\left(\left|(D)\sum_{i=1}^{n} f_{0_\pi}(\xi^i)\prod_{r=1}^{m} B(I_{i_r}) - IM(f)\right|^2\right) \\
&< \frac{1}{m!}\sum_{\pi \in S_m} \epsilon \\
&= \epsilon
\end{aligned}$$

thereby completing our proof. □

From Theorem 5.5, we may assume without loss of generality that the integrand of the Multiple Itô–McShane integral is symmetric (see part (iii) of Definition 5.4).

Lemma 5.2. *Let A be a set of λ^m-measure zero in T^m. Suppose that $A \cap B = \phi$, where B is the diagonal set in T^m. Then 1_A is both McShane (or Lebesgue) integrable and Multiple Itô–McShane integrable on T^m and $IM(1_A) = 0$.*

Proof. We remark that the condition $A \cap B$ is empty is necessary. It is sufficient to consider the case where A lies entirely in one of the set G contiguous to the diagonal. Let $\epsilon > 0$. Then there exists an open set $O \subset G$ such that $A \subset O$ and $O \cap B = \phi$ with $\lambda^m(O) < \epsilon$. Now we shall define $\delta(\xi) > 0$ on T^m. If $\xi \in A$, define $\delta(\xi) > 0$ such that $I \subset O$ whenever (I, ξ) is δ-fine. It is possible since O is open. If $\xi \notin A$, then $\delta(\xi)$ can take any positive value. Let $D = \{(I^{(i)}, \xi^{(i)})\}_{i=1}^n$ be any standard δ-fine division of T^m. Then

$$E\left(\left|(D)\sum_{i=1}^n 1_A(\xi^{(i)})\prod_{k=1}^m B(I_k^{(i)})\right|^2\right) = E\left((D)\sum_{i=1}^n 1_A^2(\xi^{(i)})\prod_{k=1}^m B^2(I_k^{(i)})\right)$$

$$= \sum_{\xi^{(i)}\in A}\prod_{k=1}^m \lambda(I_k^{(i)}) < \lambda^m(O) < \epsilon.$$

Hence $IM(1_A) = 0$, and our proof is complete. □

Lemma 5.3 (Henstock's Lemma). *Let f be Multiple Itô–McShane integrable on T^m and $F(I) = IM(f(I))$ for any subinterval $I \subset T^m$, where $IM(f(I)) = IM(f1_I)$. Then for each $\epsilon > 0$, there exists a positive function δ on T^m such that whenever $D = \{(I^i, \xi_i)\}_{i=1}^n$ is a δ-fine partial division of T^m, we have*

$$E\left(\left|\sum_{i=1}^n (f_0(\xi_i)B(I^i) - F(I^i))\right|^2\right) < \epsilon.$$

Proof. Let $\epsilon > 0$. Then there exists a positive function δ on T^m such that whenever $\mathcal{D}$ is a McShane δ-fine standard division of T^m, we have

$$E\left(|S(f_0, \mathcal{D}, \delta) - IM(f)|^2\right) < \epsilon.$$

Let $D = \{(I^i, \xi_i)\}_{i=1}^n$ be a McShane δ-fine partial division of T^m. Then the set $T^m \setminus \bigcup_{i=1}^n I^i$ consists of a finite number of disjoint left-open intervals, say $J^i, i = 1, 2, \ldots, m$. For $\eta > 0$, the integrability of f over each J^i implies that there exists a δ-fine division D_i of J^i such that

$$E\left(\left|S(f_0, D_i, \delta) - IM(f(J^i))\right|^2\right) < \frac{\eta}{m^2}.$$

Then $D' = D \cup D_1 \cup \cdots \cup D_m$ is a McShane δ-fine division of T^m and

$$S(f_0, D', \delta) - IM(f) = S(f_0, D, \delta) - \sum_{i=1}^n IM(f(I^i))$$

$$+ \sum_{i=1}^m \left(S(f_0, D_i, \delta) - IM(f(J^i))\right).$$

Thus

$$E\left(\left|S(f_0, D, \delta) - \sum_{i=1}^n IM(f(I^i))\right|^2\right) < \left(\epsilon^{\frac{1}{2}} + \sum_{i=1}^m \left(\frac{\eta}{m^2}\right)^{\frac{1}{2}}\right)^2 = \left(\epsilon^{\frac{1}{2}} + \eta^{\frac{1}{2}}\right)^2.$$

Observe that η is arbitrary. Therefore

$$E\left(\left|S(f_0, D, \delta) - \sum_{i=1}^{n} IM(f(I^i))\right|^2\right) < \epsilon,$$

i.e., $E\left|\sum_{i=1}^{n}\left(f_0(\xi_i)B(I^i) - F(I^i)\right)\right|^2 < \epsilon$, thereby completing the proof. □

5.3 Stochastic Properties of the Integral

Recall that if $I = \prod_{i=1}^{m} I_k$ and $J = \prod_{i=1}^{m} J_k$ are left-open intervals from T^m, then we use the notation

$$B(I) = \prod_{i=1}^{m} B(I_k)$$

and similar notation for $B(J)$. We shall next state the stochastic properties of Multiple Itô–McShane integral.

Lemma 5.4. *Let f be Multiple Itô–McShane integrable on T^m, and let $F(I)$ be the Multiple Itô–McShane integral of f on the subinterval $I \subset T^m$. Let I and J be two disjoint intervals from T^m in the same contiguous set G_π for some $\pi \in S_m$, and $c \in \mathbb{R}$ be a constant. Then*

(i) *$E[B(I)B(J)] = 0$;*
(ii) *F has the orthogonal increment property, that is, $E[F(I)F(J)] = 0$;*
(iii) *$E[B(I)F(J)] = 0$; and*
(iv) *$E[(cB(I) - F(I))(cB(J) - F(J))] = 0$.*

Proof. If $I = \prod_{i=1}^{m} I_i$ and $J = \prod_{k=1}^{m} J_k$, then it is clear that there exists an $I_i, i = 1, 2, \cdots, m$ such that $I_i \cap J_k$ is empty for all $k = 1, 2, \ldots, m$. Then there exist one I_l and one J_h, for convenience, say $I_l = I_1$ and $J_h = J_1$, such that $I_1 \cap J_1 = \emptyset$. Then

$$E[B(I)B(J)] = E\left(\prod_{i=2}^{m} B(I_i) \prod_{i=2}^{m} B(J_i)\right) E(B(I_1)B(J_1)).$$

Thus $E[B(I)B(J)] = 0$ since $I_1 \cap J_1 = \emptyset$. Hence (i) holds.

We remark that (i) only holds true for I and J in the same G_π. For example, in $T^2[0,1]$, $I = \left[\frac{1}{2}, \frac{3}{4}\right] \times \left[0, \frac{1}{4}\right]$, $J = \left[0, \frac{1}{4}\right] \times \left[\frac{1}{2}, \frac{3}{4}\right]$ are in different G_π, $E[B(I)B(J)] \neq 0$.

The proof of parts (ii), (iii) and (iv) follow from part (i) and similar to Lemma 2.2 in Chapter 2, hence omitted. □

Lemma 5.5. *Let f be Multiple Itô–McShane integrable on T^m with $F(I)$ to denote the integral on the interval I. Let a positive function δ on T^m be given and $D = \{(I,\xi)\}$ be a standard δ-fine partial division of T^m. Assume also that $I \in G_\pi$ whenever $\xi \in G_\pi$. Then*

(i) $E\left(|\sum f 1_{G_\pi}(\xi)B(I)|^2\right) = \sum f^2 1_{G\pi}(\xi)|I|$; *and*

(ii) $E\left(|\sum f 1_{G_\pi}(\xi)B(I) - F(I)|^2\right) = E\left(\sum |f 1_{G_\pi}(\xi)B(I) - F(I)|^2\right)$.

Proof. To prove (i), observe that

$$\begin{aligned} E\left(\left|\sum f 1_{G_\pi}(\xi)B(I)\right|^2\right) &= \sum E\left(f^2 1_{G\pi}(\xi)B(I)^2\right) \qquad \text{by Lemma 5.4(i)} \\ &= \sum f^2(\xi)|I|. \end{aligned}$$

To prove (ii): by Lemma 5.4(iv) we have

$$E\left(\left|\sum (f 1_{G_\pi}(\xi)B(I) - F(I)\right|^2\right) = E\left(\sum |(f 1_{G_\pi}(\xi)B(I) - F(I)|^2\right)$$

thereby completing our proof. □

Remark 5.2. It can be seen from the Henstock integration theory, see, e.g., [Henstock (1988); Lee (1989); Lee T.Y. (2011); McShane (1984)], that $f \in L^2(T^m, \lambda^m)$ if and only if f is McShane integrable on T^m, i.e., given any $\epsilon > 0$, there exists a positive function δ such that for any δ-fine division of T^m, denote by $D = \{(I,x)\}$, we have

$$\left|(D)\sum f^2(x)\lambda^m(I) - \int_{T^m} f^2 d\lambda^m\right| < \epsilon.$$

Also since the diagonal is a set of Lebesgue-measure zero, then f is integrable if and only if f_0 is integrable there.

Theorem 5.6 (Itô-isometry). *Let $f : T^m \to \mathbb{R}$. Suppose that f is Multiple Itô–McShane integrable. Then f^2 is Lebesgue integrable in T^m and for each G_π,*

$$E\left(IM(f 1_{G_\pi})^2\right) = (L)\int_{T^m} f^2 1_{G_\pi}(t)dt. \tag{5.1}$$

Proof. We have

$$\begin{aligned} E(IM(f 1_{G_\pi})) &= \lim_{n\to\infty} E\left(|S(f 1_{G_\pi}, D_n, \delta_n)|^2\right) \\ &= \lim_{n\to\infty} E\left(\sum f 1_{G_\pi}(x)B(I)\right)^2 \\ &= \lim_{n\to\infty} \sum f^2 1_{G_\pi}(x)|I| \qquad \text{by Lemma 5.5(i).} \end{aligned}$$

The Riemann sum $\sum f^2 1_{G_\pi}|I|$ converges to $\int_{T^m} f^2 1_{G_\pi} d\lambda^m$, and hence

$$\int_{T^m} f^2 1_{G_\pi} d\lambda^m = E\left(IM(f 1_{G_\pi})^2\right).$$

For each G_π, $f^2 1_{G_\pi}$ is Lebesgue integrable on T^m. Hence f^2 is Lebesgue integrable on T^m. The proof is complete. □

Subsequently, we shall let $IM(f)$ to denote the Multiple Itô–McShane integral of f on T^m; if I is an interval from T^m, we shall let $IM(f)(I)$ or $IM(f(I))$ denote the Multiple Itô–McShane integral of f on I.

Theorem 5.7. *Let f and g be Multiple Itô–McShane integrable on T^m and $\{I_i\}$ be a finite disjoint collection of intervals from G_π. Then*

(i) $E\left(IM(f)\right) = 0$;
(ii) $E\Big(\sum\limits_i IM(f)(I_i)\Big)^2 = \sum\limits_i E(IM(f)(I_i))^2$;
(iii) $E\left(IM(f 1_{G_\pi}) IM(g 1_{G_\pi})\right) = \int_{T^m} f g 1_{G_\pi}(t) d\lambda^m$;
(iv) $E\left(IM(f 1_{G_\pi})\right)^2 = \int_{T^m} f^2 1_{G_\pi}(t) d\lambda^m$; *and*
(v) $E\left(IM(f)\right)^2 = m! \int_{T^m} (\tilde{f})^2 d\lambda^m \le m! \int_{T^m} f^2 d\lambda^m$.

Proof. It is easy to see by the orthogonal property of Brownian motion that

$$E\left(S(f_0, \delta_n, D_n)\right) = 0.$$

Hence by Lemma 5.3,

$$E(IM(f)) = \lim_{n\to\infty} E\left(S(f_0, \delta_n, D_n)\right) = 0$$

thereby completing the proof of (i). Part (ii) is a direct consequence of Lemma 5.5. Part (iii) is also proved similarly as (i) since

$$\begin{aligned} E(S(f_0 1_{G_\pi}, \delta_n, D_n) S(g_0 1_{G_\pi}, \delta_n, D_n) &= E\Big((D_n)\sum f_0(x) g_0(x) 1_{G_\pi}(x) B^2(J)\Big) \\ &= \sum f_0(x) g_0(x) 1_{G_\pi}(x)|J| \end{aligned}$$

which tends to $\int_{T^m} f_0 g_0 1_{G_\pi}(x) d\lambda^m$. (iv) is a consequence of (iii).

Now we shall prove (v).

$$\begin{aligned} E(IM(f))^2 &= E(IM(\tilde{f}))^2 \\ &= E\left(\sum_{\pi\in S_m}\left(IM\left(\tilde{f}1_{G_\pi}\right)\right)\right)^2 \\ &= E\left(m!\left(IM\left(\tilde{f}1_{G_\pi}\right)\right)\right)^2 \\ &= (m!)^2\int_{T^m}(\tilde{f})^2 1_{G_\pi}d\lambda^m \\ &= m!\int_{T^m}(\tilde{f})^2 d\lambda^m. \end{aligned}$$

For the second part of (v),

$$\begin{aligned} \left(\int_{T^m}(\tilde{f})^2 d\lambda^m\right)^{\frac{1}{2}} &= \frac{1}{m!}\left(\int_{T^m}\left(\sum_{\pi\in S_m} f_\pi\right)^2 d\lambda^m\right)^{\frac{1}{2}} \\ &\leq \frac{1}{m!}\sum_{\pi\in S_m}\left(\int_{T^m} f_\pi^2 d\lambda^m\right)^{\frac{1}{2}} \\ &= \frac{1}{m!}(m!)\left(\int_{T^m} f^2 d\lambda^m\right)^{\frac{1}{2}} \\ &\leq \left(\int_{T^m} f^2 d\lambda^m\right)^{\frac{1}{2}}. \end{aligned}$$

So, the proof is complete. □

5.4 Integrable Functions

In this section, we characterise the class of all Multiple Itô–McShane integrable functions on T^m. We shall show that f is Multiple Itô–McShane integrable on T^m if and only if $f \in L^2(T^m, \lambda^m)$, that is,

$$(L)\int_{T^m} f^2 d\lambda^m < \infty.$$

Let $f \in L^2(T^m, \lambda^m)$, we define

$$\|f\|_m = \left(\int_{T^m} f^2 d\lambda^m\right)^{\frac{1}{2}}.$$

Given a positive function δ, we shall let

$$\tilde{S}(f, \delta, D) = (D)\sum f^2(x)\lambda^m(I)$$

where D is a δ-fine division of T^m.

Lemma 5.6. *Let $f1_{G_\pi}$ be a function on T^m. Then*

$$E\left(|S(f1_{G_\pi}, \delta, D)|^2\right) = \tilde{S}(f1_{G_\pi}, \delta, D).$$

Proof. This is just a re-iteration of part (i) of Lemma 5.5, hence the proof is omitted. □

As in Theorem 2.11 or Theorem 5.3, we have the following theorem:

Theorem 5.8. *Let f be a function defined on T^m. Then $f1_{G_\pi} \in L^2(T^m, \lambda^m)$ if and only if there exists a real number A, a decreasing sequence $\{\delta_n\}$ of positive functions on T^m such that for any sequence of δ_n-fine division D_n of T^m, we have*

$$\lim_{n\to\infty} \left|\tilde{S}(f1_{G_\pi}, \delta_n, D_n) - A\right| = 0.$$

Theorem 5.9. *Let f be a function on T^m. Then $f1_{G_\pi}$ is Multiple Itô–McShane integrable on T^m if and only if $f1_{G_\pi} \in L^2(T^m, \lambda^m)$. Furthermore,*

$$E(IM(f1_{G_\pi})^2) = (L)\int_{T^m} f^2 1_{G_\pi} d\lambda^m.$$

Proof. By Theorem 5.6, if $f1_{G_\pi}$ is Multiple Itô–McShane integrable, then $f1_{G_\pi}$ is Lebesgue integrable there. We just need to prove the converse. Suppose $f1_{G_\pi}$ is Lebesgue integrable. By Theorem 5.8,

$$(L)\int_{T^m} f^2 1_{G_\pi} d\lambda^m = \lim_{n\to\infty} \tilde{S}(f1_{G_\pi}, \delta_n, D_n) = \lim_{n\to\infty} E\left(|S(f1_{G_\pi}, D_n, \delta_n)|^2\right)$$

and by Theorem 5.3, we have that $f1_{G_\pi}$ is Multiple Itô–McShane integrable on T^m, thereby completing our proof. □

5.5 Convergence Theorem

In this section, we establish Convergence Theorems that pertain to the Multiple Itô–McShane integral. For the convenience of the readers to follow and to establish the generalization of results from Chapter 2, the results of Convergence Theorems are established parallel to the results in Chapter 2 for the Itô integral.

In this section, we consider the integral of f over a contiguous set G_π of T^m. We may assume that f vanishes over $T^m \backslash G_\pi$, or we consider the integral of $f1_{G_\pi}$ over T^m in our subsequent consideration.

We recall the notation that if f is Multiple Itô–McShane integrable on T^m, we let $IM(f)$ denote the integral on T^m. For any subinterval

$J \subset T^m$, we let $IM(f)(J)$ or $IM(f(J))$ denote the integral of f over the subinterval J.

Lemma 5.7. *Let $\{f_n\}$ be a sequence of Multiple Itô–McShane integrable functions defined on $T^m \cap G_\pi$. Suppose that*

$$E\left(IM(f_n 1_{G_\pi}) - IM(f_m 1_{G_\pi}))\right)^2 \to 0$$

as $n, m \to \infty$. Then there exists $A(I) \in L^2(\Omega)$ for every subinterval I of T^m with the following properties:

(i) *for every $\epsilon > 0$, there exists a positive integer N such that for every finite collection of disjoint left-open subintervals $\{J^i\}_{i=1}^p$ of $T^m \cap G_\pi$, we have*

$$E\left(\sum_{i=1}^{p} \left(IM(f_n(J^i)) - A(J^i)\right)\right)^2 = \sum_{i=1}^{p} E\left(IM(f_n(J^i) - A(J^i)\right)^2 < \epsilon$$

whenever $n \geq N$; and

(ii) *for each $\epsilon > 0$, there exists a subsequence $\{f_{n_k}\}$ of $\{f_n\}$ such that for each fixed k, for every finite collection of disjoint left-open subintervals $\{J^i\}_{i=1}^p$ of $T^m \cap G_\pi$, we have*

$$E\left(\left|\sum_{i=1}^{p} \left(IM(f_{n_k}(J^i) - A(J^i)\right)\right|^2\right)$$

$$= \sum_{i=1}^{p} E\left(\left|IM(f_{n_k}(J^i) - A(J^i)\right|^2\right) < \frac{\epsilon}{2^k}.$$

Proof. (i) For each subinterval I of $T^m \cap G_\pi$, we get

$$E\left(IM(f_n(I)) - IM(f_m(I))\right)^2 \to 0$$

as $n, m \to \infty$. Hence by the completeness of $L^2(\Omega)$, for each $I \subset T^m \cap G_\pi$, there exists an $A(I) \in L^2(\Omega)$ such that

$$E\left(IM(f_n(I)) - A(I)\right)^2 \to 0$$

as $n \to \infty$.

On the other hand, for every $\epsilon > 0$, there exists a positive integer N such that whenever $m, n \geq N$, we have

$$E\left(IM(f_n 1_{G_\pi}) - IM(f_m 1_{G_\pi})\right)^2 = E\left(IM(f_n) - IM(f_m)\right)^2 < \epsilon.$$

Recall that we assume that f vanishes over $T^m \setminus G_\pi$. Observe that, by Theorem 5.7(ii), for every finite collection of disjoint left-open subintervals $\{I^i\}_{i=1}^p$ of $T^m \cap G_\pi$, we have

$$E\left(\sum_{i=1}^{p} IM(f_n(I^i)) - IM(f_m(I^i))\right)^2 = \sum_{i=1}^{p} E\left(IM(f_n(I^i)) - IM(f_m(I^i))\right)^2 \leq E\left(IM(f_n) - IM(f_m)\right)^2$$

for any positive integer n, m. Hence

$$E\left(\sum_{i=1}^{p}\left(IM(f_n(I^i)) - IM(f_m(I^i))\right)\right)^2 < \epsilon$$

for all $m, n \geq N$ and any disjoint interval $\{I^i\}_{i=1}^p$ of $T^m \cap G_\pi$. By letting $m \to \infty$, we get

$$E\left(\sum_{i=1}^{p}\left(IM(f_n(I^i) - A(I^i)\right)\right)^2 < \epsilon$$

for all $n \geq N$, hence completing the proof of (i).

(ii) Apply (i) with ϵ replaced by $\frac{\epsilon}{2^k}$, $k = 1, 2, \ldots$. We can choose a subsequence $\{f_{n_k}\}$ of $\{f_n\}$ with the required property. □

Theorem 5.10 (Mean Convergence Theorem). *Let $\{f_n\}$ be a sequence of Multiple Itô–McShane integrable functions defined on $T^m \cap G_\pi$. Suppose that*

(i) *$f_n(x) \to f(x)$ almost everywhere on $T^m \cap G_\pi$ as $n \to \infty$; and*
(ii) *$E\left((IM(f_n) - IM(f_m))^2\right) \to 0$ as $n, m \to \infty$.*

Then f is Multiple Itô–McShane integrable on $T^m \cap G_\pi$ and

$$E\left((IM(f_n) - IM(f))^2\right) \to 0$$

as $n \to \infty$.

Proof. Let $\epsilon > 0$ be given. By Lemma 5.7(ii), there exists $A(I) \in L^2(\Omega)$ for every subinterval $I \subset T^m \cap G_\pi$ with the following property: there exists a subsequence $\{f_{n_k}\}$ of $\{f_n\}$ such that for every finite collection of disjoint left-open subintervals $\{I^i\}_{i=1}^p$ of $T^m \cap G_\pi$ we have

$$E\left(\left|\sum_{i=1}^{p} IM(f(I^i)) - A(I^i)\right|^2\right) < \left(\frac{\epsilon}{3 \cdot 2^k}\right). \tag{5.2}$$

In the following proof, we use the above subsequence $\{f_{n_k}\}$ instead of the given sequence $\{f_n\}$. However, for the convenience of our presentation, we denote $\{f_{n_k}\}$ by $\{f_k\}$. In other words, when we apply inequality (5.2) above, each f_{n_k} is replaced by f_k, $k = 1, 2, \ldots$. Now we shall prove that f is Multiple Itô–McShane integrable to $A(T^m)$ on $T^m \setminus G_\pi$.

First, we may assume that f_n converges to f everywhere on $T^m \cap G_\pi$. For each $\xi \in T^m \cap G_\pi$, there exists a positive integer $n(\xi)$ such that

$$\left|f(\xi) - f_{n(\xi)}(\xi)\right| < \frac{\epsilon}{3}. \tag{5.3}$$

Secondly, by Henstock's Lemma, for each $n = 1, 2, \ldots$, there exists a positive function δ_n on T^m such that whenever $D' = \{(I^i, \gamma_i)\}_{i=1}^q$ is a δ_n-fine partial division of $T^m \cap G_\pi$, we have

$$E\left(\left|\sum_{i=1}^{q} f_n(\gamma_i)B(I^i) - IM(f_n(I^i))\right|^2\right) < \left(\frac{\epsilon}{3 \cdot 2^n}\right)^2. \tag{5.4}$$

Now we shall define a positive function δ on $T^m \cap G_\pi$. For each $\xi \in T^m \cap G_\pi$, define $\delta(\xi) = \delta_{n(\xi)}(\xi)$, where $n(\xi)$ is given in (5.3). Let $D = \{(I^i, \xi_i)\}_{i=1}^p$ be a δ-fine division of T^m. For convenience, denote $n(\xi_i)$ by n_i. For each positive integer $k, k = 1, 2, \cdots$, let $D_k = \{(I^i, \xi_i) \in D : n_i = k\}$. Then each D_k is a δ_k-fine partial division of $T^m \cap G_\pi$. Notice that some of the D_k's are void. Then

$$\begin{aligned}
&\left(E\left(\left|\sum_{i=1}^{p}(f(\xi_i) - f_{n_i}(\xi_i))B(I^i))\right|^2\right)\right)^{\frac{1}{2}} \\
&= \left(E\left(\sum_{i=1}^{p}(f(\xi_i) - f_{n_i}(\xi_i))^2 \left(B(I^i)\right)^2\right)\right)^{\frac{1}{2}} \\
&< \frac{\epsilon}{3}\left(E\left(\sum_{i=1}^{p}(B(I^i)^2\right)\right)^{\frac{1}{2}} \qquad \text{by (5.3)} \\
&= \frac{\epsilon}{3},
\end{aligned}$$

$$\left(E\left|\sum_{i=1}^{p}(f_{n_i}(\xi_i)B(I^i) - IM(f(I^i)))\right|^2\right)^{\frac{1}{2}}$$

$$= \left(E\left|\sum_{k=1}^{\infty}(D_k)\sum_{n_i=k}((f_{n_i}(\xi_i)B(I^i) - IM(f_{n_i}))\right|^2\right)^{\frac{1}{2}}$$

$$\le \sum_{k=1}^{\infty}\left(E\left|(D_k)\sum_{n_i=k}(f_{n_i}(\xi_i)B(I^i) - IM(f_{n_i}))\right|^2\right)^{\frac{1}{2}}$$

$$< \sum_{k=1}^{\infty}\frac{\epsilon}{3\cdot 2^k} = \frac{\epsilon}{3}$$

and

$$\left(E\left|\sum_{i=1}^{p}(IM(f_{n_i}(I^i)) - A(I^i))\right|^2\right)^{\frac{1}{2}}$$

$$= \left(E\left|\sum_{k=1}^{\infty}(D_k)\sum_{n_i=k}(IM(f_{n_i}(I^i)) - A(I^i)\right|^2\right)\Bigg)^{\frac{1}{2}}$$

$$\le \sum_{k=1}^{\infty}\left(E\left|(D_k)\sum_{n_i=k}(IM(f_{n_i}(I^i)) - A(I^i)\right|^2\right)^{\frac{1}{2}}$$

$$< \sum_{k=1}^{\infty}\frac{\epsilon}{3\cdot 2^k} = \frac{\epsilon}{3}.$$

With the above three inequalities, we have

$$\left(E\left|\sum_{i=1}^{p}f(\xi_i)B(I^i) - A(T^m)\right|^2\right)^{\frac{1}{2}} < \frac{\epsilon}{3} + \frac{\epsilon}{3} + \frac{\epsilon}{3} = \epsilon.$$

Thus f is Multiple Itô–McShane integrable to $A(T^m)$ on $T^m \cap G_\pi$ and by condition (ii), we get

$$E\left(|IM(f_n) - IM(f)|^2\right) \to 0$$

as $n \to \infty$. □

In view of the fact that f is Multiple Itô–McShane integrable on T^m if and only if f is Multiple Itô–McShane integrable on each contiguous set G_π of T^m, the following Corollary 5.2 is a direct consequence.

Corollary 5.1. *Let $\{f_n\}$ be a sequence of Multiple Itô–McShane integrable functions defined on T^m. Suppose that for each G_π,*

(i) $f_n(x) \to f(x)$ *almost everywhere on* $T^m \cap G_\pi$ *as* $n \to \infty$*; and*
(ii) $E\left((IM(f_n 1_{G_\pi}) - IM(f_m 1_{G_\pi}))^2\right) \to 0$ *as* $n, m \to \infty$.

Then f *is Multiple Itô–McShane integrable on* T^m *and*

$$E\left((IM(f_n) - IM(f))^2\right) \to 0$$

as $n \to \infty$.

Theorem 5.11 (Dominated Convergence Theorem). *Let* $\{f_n\}$ *be a sequence of Multiple Itô–McShane integrable functions defined on* $T^m \cap G_\pi$. *Suppose that*

(i) $f_n(t) \to f(t)$ *almost everywhere on* $T^m \cap G_\pi$ *as* $n \to \infty$*; and*
(ii) $|f_n(t)| \leq g(t)$ *for all* n *and for almost all* $t \in T^m \cap G_\pi$,

where g^2 *is Lebesgue integrable on* $T^m \cap G_\pi$. *Then* f *is Multiple Itô–McShane integrable on* $T^m \cap G_\pi$ *and*

$$E\left((IM(f_n) - IM(f))^2\right) \to 0$$

as $n \to \infty$.

Proof. By Theorem 5.6, each f_n is square Lebesgue integrable on $T^m \cap G_\pi$. By Dominated Convergence Theorem for Lebesgue integral on $T^m \cap G_\pi$, f is square Lebesgue integrable on $T^m \cap G_\pi$ and $(L)\int_{T^m}(f_n - f)^2 1_{G_\pi} \to 0$ as $n \to \infty$. Hence $(L)\int_{T^m}(f_n - f_m)^2 1_{G_\pi} \to 0$ as $m, n \to \infty$. By Itô-isometric property (Theorem 5.6),

$$E\left(IM(f_n) - IM(f_m)\right)^2 \to 0$$

as $n, m \to \infty$. Now apply Theorem 5.10 (Mean Convergence Theorem) to $\{f_n\}$, we get the required result. □

Corollary 5.2. *Let* $\{f_n\}$ *be a sequence of Multiple Itô–McShane integrable functions defined on* T^m. *Suppose for each* G_π,

(i) $f_n(x) \to f(x)$ *almost everywhere on* $T^m \cap G_\pi$ *as* $n \to \infty$*; and*
(ii) $|f_n(x)| \leq g(x)$ *for all* n *and almost everywhere on* $T^m \cap G_\pi$,

where g *is Multiple Itô–McShane integrable on* T^m, *then* f *is Multiple Itô–McShane integrable on* T^m *and*

$$E\left((IM(f_n) - IM(f))^2\right) \to 0$$

as $n \to \infty$.

5.6 Classical Multiple Itô–Wiener Integral

In this section, we shall review the classical construction of the (classical) Multiple Itô–Wiener integral.

Definition 5.5. Let $\{I_1, I_2, \ldots, I_n\}$ be a collection of left-open subintervals which form a partition of $(0,1]$, i.e., the intervals $I_1, I_2, \ldots, I_n$ are disjoint and $\bigcup_{k=1}^{n} I_k = (0,1]$. An *elementary function* on T^m is a function $g : T^m \to \mathbb{R}$ that can be expressed in the form

$$g = \sum_{i_1, i_2, \ldots, i_m = 1}^{n} a_{i_1, i_2, \ldots, i_m} 1_{I_{i_1} \times I_{i_2} \times \cdots \times I_{i_m}} \tag{5.5}$$

where $\{I_{i_1}, I_{i_2}, \ldots, I_{i_m}\}$ is a subset from $\{I_1, I_2, \ldots, I_n\}$ and $a_{i, i_2, \ldots, i_m} = 0$ if any two of the indices $i_1, i_2, \ldots, i_m$ are equal.

Note that the definition of an elementary function shows that g vanishes on the diagonal of T^m, that is, if $t = (t_1, t_2, \ldots, t_n) \in T^m$ such that $t_i = t_j$ for some $i \neq j$, then $g(t) = 0$.

The Multiple Itô–Wiener integral of an elementary function g is defined by

$$IW(g) = \sum_{i_1, i_2, \ldots, i_m = 1}^{n} a_{i_1, i_2, \ldots, i_m} \prod_{j=1}^{m} B(I_{i_j}),$$

where B is the Brownian motion and $B(I)$ denotes $B_b - B_a$ if $I = (a, b]$. It is known that

$$E\left(IW(g)\right)^2 \leq m! \|g\|_m^2,$$

see [Itô (1951); Norin (1996), p. 8] or Theorem 5.7(v).

Let $f \in L^2(T^m, \lambda^m)$. Then there exists a sequence $\{f_n\}$ of elementary functions on T^m such that $\lim_{n\to\infty} \|f_n - f\|_m = 0$. On the other hand,

$$E\left(IW(f_p - f_q)\right)^2 \leq m! \|f_p - f_q\|_m^2.$$

Hence $\{IW(f_n)\}$ is a Cauchy sequence in $L^2(\Omega)$. By completeness, the limit exists and hence the Multiple Itô–Wiener integral $IW(f)$ of f is defined as

$$\lim_{n\to\infty} E\left((IW(f_n) - IW(f)\right)^2 = 0,$$

see [Itô (1951)]. This process is similar to the construction of Itô integral with respect to Brownian motion in one-dimensional space.

Theorem 5.12. *Let g be an elementary function on T^m in the form* (5.5). *Then g is Multiple Itô–McShane integrable on T^m and*

$$IM(g) = \sum_{i_1, i_2, \ldots, i_m = 1}^{n} a_{i_1, i_2, \ldots, i_m} \prod_{k=1}^{m} B(I_{i_k}).$$

Proof. It is sufficient to prove the special case when $g = 1_{I_1 \times I_2 \times \cdots \times I_m}$, where I_i, $i = 1, 2, \ldots, m$, are disjoint left-open subintervals of $[0,1]$. Let $I = I_1 \times I_2 \times \cdots \times I_m$ and ∂I be the boundary of I. It is clear by definition that g does not vanish only on one of the set G contiguous to the diagonal. Denote $\prod_{k=1}^{m} B(I_k)$ by $B(I)$. By Lemma 5.2, we need only to consider $g = 1_{I \backslash \partial I}$ since ∂I is a set of measure zero. Let $\epsilon > 0$. There exists an open set $O \supset \partial I$ and $G \supset O$ such that $\lambda^m(O) < \epsilon$. Now we shall define $\delta(\xi) > 0$ on T^m in the following way: If $\xi \in I \backslash \partial I$, define the positive function δ to be such that $J \subset I \backslash \partial I$ whenever (J, ξ) is δ-fine. If $\xi \in \partial I$, define $\delta(\xi) > 0$ such that $J \subset O$ whenever (J, ξ) is δ-fine. If $\xi \notin I$, then $\delta(\xi)$ can take any positive value. Let $D = \{(J^{(i)}, \xi^{(i)})\}_{i=1}^{n}$ be a δ-fine division of T^m, where $J^{(i)} = J_1^{(i)} \times J_2^{(i)} \times \cdots \times J_m^{(i)}$. Then

$$
\begin{aligned}
& E\left(\left|(D)\sum g(\xi^{(i)})B(J^{(i)}) - B(I)\right|^2\right) \\
&= E\left(\left|(D)\sum_{\xi^{(i)} \in I\backslash \partial I} g(\xi^{(i)})B(J^{(i)}) - B(I)\right|^2\right) \\
&= E\left(\left|(D)\sum_{\xi^{(i)} \in I\backslash \partial I} B(J^{(i)}) - B(I)\right|^2\right) \\
&\le 2E\left(\left|(D)\sum_{\xi^{(i)} \in I} B(J^{(i)}) - B(I)\right|^2\right) + 2E\left(\left|(D)\sum_{\xi^{(i)} \in \partial I} B(J^{(i)})\right|^2\right) \\
&< 2E\left(\left|(D)\sum_{\xi^{(i)} \in I} B(J^{(i)}) - B(I)\right|^2\right) + 2\epsilon \quad \text{by Lemma 5.5 and } \lambda^m(O) < \epsilon \\
&\le 4E\left(\left|(D)\sum_{\xi^{(i)} \in I} \left[B(J^{(i)}) - B(I \cap J^{(i)})\right]\right|^2\right) \\
&\quad + 4E\left(\left|(D)\sum_{\xi^{(i)} \in I} B(I \backslash J^{(i)})\right|^2\right) + 2\epsilon \\
&< 4\epsilon + 0 + 2\epsilon, \qquad \text{by Lemma 5.5 and } \lambda^m(O) < \epsilon \\
&= 6\epsilon.
\end{aligned}
$$

Hence $IM(f) = B(I) = \prod_{k=1}^{m} B(I_k)$. From the classical definition,

$$IW(f) = \prod_{k=1}^{m} B(I_k).$$

Therefore f is Multiple Itô–McShane integrable with $IM(f) = IW(f)$. □

We are now ready to prove the equivalence of the classical Multiple Itô–Wiener integral and the Multiple Itô–McShane integral.

Theorem 5.13 (Equivalence Theorem). *Let $f \in L^2(T^m, \lambda^m)$. Then f is Multiple Itô–McShane integrable on T^m and $IM(f) = IW(f)$.*

Proof. By Theorem 5.12, the result of Theorem 5.13 holds true for any elementary functions of the form (5.5). Let $f \in L^2(T^m, \lambda^m)$. By the definition of the Multiple Itô–Wiener integral, there exists a sequence of elementary functions $\{f_n\}$ such that $f^{(n)}$ converges to f a.s. on T^m and $\|f^{(n)} - f\|_m \to 0$ as $n \to \infty$. Furthermore $E\left(IW(f^{(n)} - f)\right)^2 \leq m!\, \|f^{(n)} - f\|_m$ for all positive integer n, and

$$\lim_{n\to\infty} E\left(IW(f^{(n)}) - IW(f)\right)^2 = 0.$$

On the other hand,

$$E\left(IM\left(f^{(n)}\right) - IM\left(f^{(m)}\right)\right)^2 = E\left(IW\left(f^{(n)}\right) - IW\left(f^{(m)}\right)\right)^2 \to 0$$

as $m, n \to \infty$.

By Mean Convergence Theorem, Theorem 5.10 and its Corollary 5.1, f is Multiple Itô–McShane integrable on T^m and $E\left(IM\left(f^{(n)}\right) - IM\left(f\right)\right)^2 \to 0$ as $n \to \infty$. Hence $IM(f) = IW(f)$. □

5.7 Notes and Remarks

Results and their proofs in this chapter using non-uniform Riemann approach can be found in [Toh (2001); Toh and Chew (2003–04, 2004–05, 2005)].

We use δ-fine full divisions in this chapter since δ-fine interval-point pairs are not belated δ-fine. Many ideas in this chapter are similar to ideas in Chapter 2. Some proofs are easier since full divisions are used and integrands are deterministic.

Chapter 6

Fubini's Theorem and Hu–Meyer's Theorem

In this chapter, first we shall define Multiple Stochastic integrals. The integrators may not be Brownian Motions in one-dimensional space. We shall derive the Henstock–Fubini's Theorem for Multiple Stochastic Integrals based on the Henstock approach and show that the Iterated Multiple Integral Formula is a direct consequence of Henstock–Fubini's Theorem.

In Chapter 5, Section 5.1, we saw that the Itô–McShane integral (see Definition 5.2) does not take into account the values of the integrand f on the diagonal (and replace the integrand f by f_0, which vanishes on the diagonal). As we have seen in the preceding chapter, it turns out that the Multiple Itô–McShane integral is identical to the classical Multiple Itô–Wiener Integral. In this chapter we shall study the integral on the diagonal and offer a proof of the classical Hu–Meyer Theorem.

6.1 Fubini's Theorem

6.1.1 *Setting*

Let $(\Omega, \mathcal{F}, P)$ be a complete probability space, $\mathbb{R} = (-\infty, \infty)$, $T = [a, b]$ and $T^m = [a, b] \times [a, b] \times \cdots \times [a, b]$, that is, m copies of $[a, b]$. An interval $I \subset T^m$ is said to be *left-open* if $I = \prod_{i=1}^{m}(a_i, b_i] \subset T^m$, where each $(a_i, b_i]$ is a left-open interval in T. Let $\mathcal{G}_m$ be the collection of all left-open intervals in T^m.

It is noted that T^m can be decomposed into two parts: the diagonal part $\mathcal{D}$ consisting of

$$\mathcal{D} = \{(x_1, x_2, x_3, \ldots, x_m) \in T^m : x_i = x_j \text{ for some } i \neq j\}$$

and the non-diagonal part $\mathcal{D}^c$ which consists of

$$\mathcal{D}^c = \{(x_1, x_2, \ldots, x_m) : x_i \neq x_j \text{ whenever } i \neq j\}.$$

In addition, the diagonal part can be decomposed into $m!$ connected sets contiguous to the diagonal in the following way:

Let S_m be the set of all permutation of m distinct objects. Hence there are $m!$ elements in S_m. For each $\pi \in S_m$, define

$$G_\pi = \{(x_1, x_2, \ldots, x_m) : x_{\pi(1)} < x_{\pi(2)} < x_{\pi(3)} < \cdots < x_{\pi(m)}\}.$$

Note that each set G_π is open in T^m. The $m!$ contiguous sets are disconnected from one another and disjoint from the diagonal.

For each $\pi \in S_m$, we shall denote the projection of G_π to T^p, where $p \leq m$, by $\mathrm{Proj}_p(G_\pi)$, that is,

$$\mathrm{Proj_p(G_\pi)} = \{(\mathrm{x_1, x_2, \ldots, x_p}) \in \mathbb{R}^\mathrm{p} : (\mathrm{x_1, x_2, \ldots, x_p, x_{p+1}, \ldots, x_m}) \in \mathrm{G_\pi}\}.$$

Let $I = \prod_{i=1}^{p} I_i = I_1 \times I_2 \times \cdots \times I_p \in \mathcal{G}_p$, where each $I_i \subset \mathbb{R}$ is of the form $(a_i, b_i]$. The interval I is said to be *non-diagonal* if $I \subset \mathrm{Proj_p(G_\pi)}$ for some $\pi \in S_m$.

Definition 6.1. Let $X : \mathcal{G}_p \times \Omega \to \mathbb{R}$ and $Y : \mathcal{G}_q \times \Omega \to \mathbb{R}$ such that $E(X^2(I)) < \infty$ and $E(Y^2(J)) < \infty$ for all $I \in \mathcal{G}_p$, $J \in \mathcal{G}_q$ and $p + q = m$. Then X and Y are said to be *uncorrelated of second order* if

$$\begin{aligned} &E\left(X(I^{(1)})X(I^{(2)})Y(J^{(1)})Y(J^{(2)})\right) \\ &= E\left(X(I^{(1)})X(I^{(2)})\right) E\left(Y(J^{(1)})Y(J^{(2)})\right) \end{aligned}$$

whenever $I^{(i)} \times J^{(j)} \subset G_\pi$ for some fixed $\pi \in S_m$, $i, j = 1, 2$.

Definition 6.2. Let $X : \mathcal{G}_p \times \Omega \to \mathbb{R}$. Then X is said to satisfy the *orthogonal property* if $E\left(X(J)X(K)\right) = 0$ for all disjoint pair of intervals $J, K \in \mathcal{G}_p$ and which are in $\mathrm{Proj_p(G_\pi)}$ for some fixed $\pi \in S_m$.

6.1.2 *Multiple stochastic integral*

We begin this section by defining the non-uniform division of T^m that we shall take.

Let δ be a positive function defined on T^m, $\xi = (\xi_1, \xi_2, \ldots, \xi_m) \in T^m$ and $I = \prod_{i=1}^{m} I_i$ be an interval of T^m. An *interval-point pair* (I, ξ) is said to be δ-fine if $I_k \subset (\xi_k - \delta(\xi), \xi_k + \delta(\xi))$ for each $k = 1, 2, \ldots, m$.

Note that ξ_k may or may not be in I_k for each $k = 1, 2, \ldots, m$. A finite collection D of δ-fine interval-point pairs $\{(I^{(i)}, \xi^{(i)})\}_{i=1}^{n}$ is said to be a δ-fine division of T^m if

(i) $I^{(i)}$, $i = 1, 2, \ldots, n$, are disjoint left-open intervals of T; and
(ii) $\bigcup_{i=1}^{n} I^{(i)} = (0, 1]^m$.

We remark that for any given positive function δ on T^m, a δ-fine division of T^m exists, by Cousin's Lemma, see for examples [Lee T.Y. (2011), p. 25; Henstock (1988), p. 42].

Notation 6.1. Let $f : T^m \times \Omega \to \mathbb{R}$, $Z : \mathcal{G}_m \times \Omega \to \mathbb{R}$ and δ be a given positive function on T^m. We shall denote the Riemann sum $(D)\sum_{i=1}^{n} f(\xi^{(i)}, \omega)Z(I^{(i)}, \omega)$ by $S(f, \delta, D, Z)$ if $D = \{(I^{(i)}, \xi^{(i)})\}_{i=1}^{n}$ is a δ-fine division of T^m.

Definition 6.3. The function $Z : \mathcal{G}_m \times \Omega \to \mathbb{R}$ is said to be *additive* if for each $\omega \in \Omega$,

$$Z(I \cup J, \omega) = Z(I, \omega) + Z(J, \omega)$$

whenever $I, J \in \mathcal{G}_m$ and $I \cup J \in \mathcal{G}_m$ while I and J are disjoint.

Definition 6.4. A function $f : T^m \times \Omega \to \mathbb{R}$ is said to be *Multiple Stochastic integrable* to a function $M(f)$ with respect to an additive function $Z : \mathcal{G}_m \times \Omega \to \mathbb{R}$ on the interval T^m if for every $\epsilon > 0$, there exists a positive function δ on T^m such that

$$E\left(|S(f, \delta, D, Z) - M(f)|^2\right) < \epsilon$$

whenever $D = \{(I^{(i)}, x^{(i)})\}_{i=1}^{n}$ is a δ-fine division of T^m.

Here Z is the integrator and f is the integrand. $M(f)$ is the primitive function of f with respect to Z on T^m.

We remark that when $Z(I, \omega) = B(I, \omega)$, where B is the Brownian motion in one-dimension, the Multiple Stochastic integral is the Multiple Itô–McShane integral defined in Chapter 5.

Definition 6.5. The integral of f over any subinterval $I \subset T^m$ is as defined in Definition 6.4 above, except that the positive function δ is defined on I instead of the entire T^m.

The following lemma is given in Chapter 5, Lemma 5.1.

Lemma 6.1. *Let δ be a positive function on T^m and $\{D_k\}$ be a finite family of δ-fine divisions of T^m. Then there exists a partition $\{A_1, A_2, \ldots, A_q\}$ of $[0,1]$ and a finite family of δ-fine divisions of T^m denoted by $\{D'_k\}$ such that each interval of any D'_k is of the form $A_{l_1} \times A_{l_2} \times \cdots \times A_{l_m}$ and each D'_k is a refinement of D_k. Furthermore,*

$$S(f, \delta, D_k) = S(f, \delta, D'_k)$$

for all k.

Definition 6.6. Let $T^m = [0,1]^m$ and $D = \{(I,\xi)\}$ be a δ-fine division of T^m. Suppose there exists a partition $\{A_1, A_2, \ldots, A_q\}$ of $[0,1]$ and every interval I in D is of the form $A_{l_1} \times A_{l_2} \times \cdots \times A_{l_m}$. Then D is called a standard δ-fine division.

In view of Lemma 6.1, we shall assume that all finite collections of δ-fine divisions of T^m that we consider in Definitions 6.4 and 6.5 are all standard divisions.

6.1.3 *Basic properties of the integral*

First we shall state some standard properties of Multiple Stochastic Integrals over T^m. For the first few standard properties, we omit the proofs as they are classical in the theory of Henstock integration theory, see for example [Henstock (1988); Lee (1989)].

Theorem 6.1. *The Multiple Stochastic Integral of f with respect to Z, if it exists, is unique.*

Theorem 6.2. *Let f and g be Multiple Stochastic Integrable with respect to Z on T^m, and let their integrals be denoted by $M(f)$ and $M(g)$ respectively, and let $k \in \mathbb{R}$ be fixed. Then kf and $f+g$ are integrable, with*

(i) $M(f+g) = M(f) + M(g)$; *and*
(ii) $M(kf) = kM(f)$.

Theorem 6.3. *Let f be Multiple Stochastic Integrable with respect to X and Y on T^m, and let their respective integrals be denoted by $M_X(f)$ and $M_Y(f)$ respectively. Then f is Multiple Stochastic Integrable over $X+Y$, and moreover,*

$$M_{X+Y}(f) = M_X(f) + M_Y(f).$$

Theorem 6.4 (Cauchy's Criterion). *A function $f : T^m \times \Omega \to \mathbb{R}$ is Multiple stochastic integrable on T^m with respect to Z if and only if given $\epsilon > 0$, there exists a positive function δ on T^m such that*

$$E\left(|S(f,\delta,D_1,Z) - S(f,\delta,D_2,Z)|^2\right) < \epsilon$$

whenever D_1, D_2 are standard δ-fine divisions of T^m.

Theorem 6.5. *A stochastic process $f : T^m \times \Omega \to \mathbb{R}$ is a Multiple integrable function on T^m with respect to Z if and only if there exists a sequence $\{\delta_n\}$ of positive functions on T^m, $n = 1, 2, \ldots$, with $\delta_{n+1}(\xi) < \delta_n(\xi)$ for all $n = 1, 2, \ldots$ such that $M(f)$ is the limit of $S(f, \delta_n, D_n, Z)$ under L_2-norm.*

Proof. For $n = 1, 2, \ldots$, there exists $\delta_n(\xi) > 0$ on T^m such that the inequality in Definition 6.4 holds with $\epsilon = \frac{1}{n}$. For each $n = 1, 2, \ldots$, fix a δ_n-fine division D_n. We may assume that $\delta_{n+1}(\xi) < \delta_n(\xi)$ for each n. Hence we have

$$\lim_{n\to\infty} E\left(|S(f, \delta_n, D_n, Z) - M(f)|^2\right) = 0$$

thereby completing the proof. □

6.1.4 *Henstock–Fubini's Theorem*

In the remaining part of Section 6.1, let $X : \mathcal{G}_p \times \Omega \to \mathbb{R}$ and $Y : \mathcal{G}_q \times \Omega \to \mathbb{R}$ and $Z(I \times J) = X(I)Y(J)$ for all $I \in \mathcal{G}_p$ and $J \in \mathcal{G}_q$. Further we shall assume that X and Y are uncorrelated of second order (see Definition 6.1) and that both X and Y have the orthogonal properties (see Definition 6.2).

Lemma 6.2. *Let $g : T^p \times \Omega \to \mathbb{R}$ be Multiple Stochastic Integrable on T^p with respect to X. Suppose we let $M(I)$ for each $I \in \mathcal{G}_p$ denote the integral of g with respect to X on each subinterval I, then*

(i) *M has the orthogonal property;*
(ii) *$E(X(J)M(I)) = 0$ whenever I and J are disjoint and in $\mathrm{Proj_p(G_\pi)}$ for a fixed $\pi \in S_m$;*
(iii) *$cX - M$ has the orthogonal property, where c is a real constant; and*
(iv) *Y and $cX - M$ are uncorrelated of second order, where c is a real constant.*

Proof. Let I and J be disjoint and in $\mathrm{Proj_p(G_\pi)}$. Given $\epsilon > 0$ choose a positive function δ to be the corresponding function as in Definition 6.5 on the intervals I and J, with the corresponding integrals $M(I)$ and $M(J)$. For such a chosen positive function δ it is clear that

$$E\left[S(g, \delta, D(I), X)S(g, \delta, D(J), X)\right] = 0,$$

where $D(I)$ and $D(J)$ are δ-fine divisions of I and J, respectively. This is because (a) X has the orthogonal property and (b) that if $\{I, J\}$ are disjoint and in $\mathrm{Proj_p(G_\pi)}$, then any pair $\{U, V\}$ of intervals such that $U \subset I$ and $V \subset J$ are disjoint and in $\mathrm{Proj_p(G_\pi)}$.

By Theorem 6.5 and the fact that $E[fh] = \lim_{n\to\infty} E[f_n h_n]$ whenever we have $\lim_{n\to\infty} E(f_n - f)^2 = 0$ and $\lim_{n\to\infty} E(h_n - h)^2 = 0$,

$$E(M(I)M(J)) = \lim_{n\to\infty} E\left[S(g, \delta_n, D(I), X)S(g, \delta_n, D(J), X)\right] = 0$$

thereby proving (i).

By using the similar reasoning as in (i) above,

$$E[X(J)M(I)] = \lim_{n\to\infty} E[X(J)S(g,\delta_n,D(I),X)] = 0$$

thereby completing the proof of (ii).

Using (i) and (ii) above, we have

$$\begin{aligned} E([cX-M](I)[cX-M](J)) &= c^2E(X(I)X(J)) + E(M(I)M(J)) \\ &\quad - cE(X(I)M(J)) - cE(X(J)M(I)) \\ &= 0, \end{aligned}$$

thereby proving (iii).

Let $I^{(i)} \times J^{(j)} \subset G_\pi$, $i,j=1,2$. It is also clear that, since X and Y are uncorrelated of second order, we have

$$\begin{aligned} &E\Big[Y(I^{(1)})Y(I^{(2)})X(J^{(1)})S(g,\delta_n,D(J^{(2)}),X)\Big] \\ &\qquad = E\left[Y(I^{(1)})Y(I^{(2)})\right] E\left[X(J^{(1)})S(g,\delta_n,D(J^{(2)}),X)\right]. \end{aligned}$$

Consequently, $cX - S_n$ and Y are uncorrelated of second order, where

$$S_n(I) = S(g,\delta_n,D(I),X)$$

for any interval I and any positive integer n, where δ_n are given in Theorem 6.5.

$$\begin{aligned} &E\Big\{Y(I^{(1)})Y(I^{(2)})\left(cX(J^{(1)}) - M(J^{(1)})\right)\left(cX(J^{(2)}) - M(J^{(2)})\right)\Big\} \\ &= \lim_{n\to\infty} E\left\{Y(I^{(1)})Y(I^{(2)})\left(cX(J^{(1)}) - S_n(J^{(1)})\right)\right. \\ &\qquad \left.\times\left(cX(J^{(2)}) - S_n(J^{(2)})\right)\right\} \\ &= \lim_{n\to\infty} E\left\{Y(I^{(1)})Y(I^{(2)})\right\} \\ &\qquad E\left\{\left(cX(J^{(1)}) - S_n(J^{(1)})\right)\left(cX(J^{(2)}) - S_n(J^{(2)})\right)\right\} \\ &= \lim_{n\to\infty} E\left\{\left(cX(J^{(1)}) - S_n(J^{(1)})\right)\left(cX(J^{(2)}) - S_n(J^{(2)})\right)\right\} \\ &\qquad \times E\left\{Y(I^{(1)})Y(I^{(2)})\right\} \\ &= E\left\{\left(cX(J^{(1)}) - M(J^{(1)})\right)\left(cX(J^{(2)}) - M(J^{(2)})\right)\right\} \\ &\qquad \times E\left\{Y(I^{(1)})Y(I^{(2)})\right\} \end{aligned}$$

thereby completing our proof of the entire lemma. □

Lemma 6.3. *Let δ be a positive function on T^p. Then*

$$E\left|(D)\sum_{i=1}^{n} a_i X(I^{(i)})\right|^2 = E\left((D)\sum_{i=1}^{n} a_i^2 X^2(I^{(i)})\right)$$

for any standard δ-fine partial division $D = \{(I^{(i)}, x^{(i)})\}_{i=1}^{n}$ for which all the intervals $I^{(i)}$ and the points $x^{(i)}$ are from $\mathrm{Proj_p}(\mathrm{G}_\pi)$. The same result applies to Y on T^q.

Proof. This follows from the fact that X has orthogonal property. □

Lemma 6.4. *Let δ be a positive function on T^p and δ^1 on T^q, and $p + q = m$. Suppose that $D^1 = \{(K^{(j)}, y^{(j)})\}_{j=1}^{r}$ is a standard δ^1-fine partial division of T^q and $D_j = \{(I_j^{(i)}, x_j^{(i)})\}_{i=1}^{n(j)}$, $j = 1, 2, \ldots, r$, are standard δ-fine partial divisions of T^p such that $K^{(j)} \times I_j^{(i)}$ are in G_π for all i, j. Then*

(i) $$\sum_{j=1}^{r}\left(E[Y^2(K^{(j)})]E\left|\sum_{i=1}^{n(j)} a_{ij}X(I_j^{(i)})\right|^2\right) = E\left(\left|\sum_{j=1}^{r}\sum_{i=1}^{n(j)} a_{ij}X(I_j^{(i)})Y(K^{(j)})\right|^2\right)$$

(ii) $$\sum_{j=1}^{r}\left(E(Y^2(K^{(j)})E\left|\sum_{i=1}^{n(j)}(a_{ij}X(I_j^{(i)}) - M(I_j^{(i)}))\right|^2\right) = E\left(\left|\sum_{j=1}^{r}\sum_{i=1}^{n(j)}(a_{ij}X(I_j^{(i)}) - M(I_j^{(i)}))Y(K^{(j)})\right|^2\right)$$

where M is as given in Lemma 6.2.

Proof. To prove part (i) of the lemma, we use the fact that X and Y are uncorrelated of second order and that each has orthogonal increment,

$$\begin{aligned}
\sum_{j=1}^{r} E[Y^2(K^{(j)})]E\left|\sum_{i=1}^{n(j)} a_{ij}X(I_j^{(i)})\right|^2 &= \sum_{j=1}^{r} E(Y^2(K^{(j)}))E\left(\sum_{i=1}^{n(j)} a_{ij}^2 X^2(I_j^{(i)})\right) \\
&= \sum_{j=1}^{r}\sum_{i=1}^{n(j)} a_{ij}^2 E\left(Y^2(K^{(j)})\right) E\left(X^2(I_j^{(i)})\right) \\
&= \sum_{j=1}^{r}\sum_{i=1}^{n(j)} a_{ij}^2 E\left(X^2(I_j^{(i)})Y^2(K^{(j)})\right) \\
&= \sum_{j=1}^{r}\sum_{i=1}^{n(j)} E\left[a_{ij}X(I_j^{(i)})Y(K^{(j)})\right]^2 \\
&= E\left(\left|\sum_{j=1}^{r}\sum_{i=1}^{n(j)} a_{ij}X(I_j^{(i)})Y(K^{(j)})\right|^2\right).
\end{aligned}$$

To prove (ii) of the same lemma, from part (iii) of Lemma 6.2 on the orthogonality of $cX - M$, we have

$$E\left(\sum_{i=1}^{n}\left(a_{ij}X(I_j^{(i)}) - M(I_j^{(i)})\right)\right)^2 = \sum_{i=1}^{n} E\left(a_{ij}X(I_j^{(i)}) - M(I_j^{(i)})\right)^2.$$

Hence, as in part (i) above,

$$\begin{aligned}
&\sum_{j=1}^{r}\left(E(Y^2(K^{(j)})E\left|\sum_{i=1}^{n}(a_{ij}X(I_j^{(i)}) - M(I_j^{(i)}))\right|^2\right) \\
&\qquad = \sum_{j=1}^{r}\left(E(Y^2(K^{(j)}))\sum_{i=1}^{n} E\left(a_{ij}X(I_j^{(i)}) - M(I_j^{(i)})\right)^2\right) \\
&\qquad = \sum_{j=1}^{r}\sum_{i=1}^{n}\left(E(Y^2(K^{(j)}))E(a_{ij}X(I_j^{(i)}) - M(I_j^{(i)}))^2\right) \\
&\qquad = \sum_{j=1}^{r}\sum_{i=1}^{n} E\left[Y^2(K^{(j)})\left(a_{ij}X(I_j^{(i)}) - M(I_j^{(i)})\right)^2\right] \\
&\qquad = E\left(\left|\sum_{j=1}^{r}\sum_{i=1}^{n}(a_{ij}X(I_j^{(i)}) - M(I_j^{(i)}))Y(K^{(j)})\right|^2\right)
\end{aligned}$$

thereby completing the proof. □

Definition 6.7. A function $f : T^q \to \mathbb{R}$ is said to be of *quadratic variation zero* with respect to Y if for any $\epsilon > 0$ there exists $\delta(\xi) > 0$ on T^q such that

$$\left| (D)\sum_{i=1}^{n} f(y^{(i)})E\left\{Y^2(I^{(i)})\right\} \right| < \epsilon$$

whenever $D = \{(I^{(i)}, y^{(i)})\}_{i=1}^{n}$ is a standard δ-fine partial division of T^q. A subset $F \subset T^q$ is said to be of *quadratic variation zero with respect to* Y if the indicator function 1_F is of quadratic variation zero with respect to Y.

We remark that a quadratic variation is also defined in Chapter 4, see Remark 4.1, Definitions 4.7 and 4.9.

Lemma 6.5. *Let $f(x) > 0$ on $F \subset T^q$, then the subset F is of quadratic variation zero if and only if $f1_F$ is of zero quadratic variation.*

Definition 6.8. The function $Y : \mathcal{G}_q \times \Omega \to \mathbb{R}$ is said to be of bounded quadratic variation if its quadratic variation, denoted by $V(Y)$, is finite and where

$$V(Y) = \inf_{\delta(\xi)>0} \sup_{D} \sum E(Y^2(I^{(i)}))$$

the supremum is taken over all δ-fine partial division D of T^q and the infimum is taken over all $\delta(\xi) > 0$ defined on T^q. For the remaining of this section, as we involve stochastic integrals of different integrators, we shall state the integrator on the prefix, for example, the stochastic integral of f with respect to X on T^m will be written as $M_m^X(f)$ to avoid confusion.

Before we prove Henstock–Fubini's Theorem, we need one fundamental lemma:

Lemma 6.6. *(See, for example [Henstock (1988), Lemma 17.1, p. 162]) Let δ be a positive function defined on $T^p \times T^q$. For each $y \in T^q$, define $\delta_{1y}(x) = \delta(x,y)/\sqrt{2}$ on T^p. For each $y \in T^q$, let*

$$D_1(y) = \{(I_y^{(i)}, x^{(i)})\}_{i=1}^{n}$$

be a McShane δ_{1y}-fine division of T^p. Define

$$\delta_2(y) = \min\left\{ \frac{\delta(x^{(i)}, y)}{\sqrt{2}} : i = 1, 2, \ldots, n(y) \right\}.$$

Then $(I_y^{(i)} \times J, (x^{(i)}, y))$ is McShane δ-fine for each i if (J, y) is δ_2-fine on T^q.

We are now ready to prove the Henstock–Fubini's Theorem. Recall that $Z(I \times J) = X(I)Y(J)$, where $I \in \mathcal{G}_p$ and $J \in \mathcal{G}_q$. The idea of the proof is obtained from the non-stochastic case in [Henstock (1988), Chapter 6]. An easy readable version can be found in [Lee and Vyborny (2000), p. 229] which written down the proof for two-dimensional Euclidean space.

Theorem 6.6 (Henstock–Fubini's Theorem). *Let $T^m = T^p \times T^q$ and $F \subset G_\pi$ for some fixed $\pi \in S_m$. Suppose $f : T^m \to \mathbb{R}$ and that $f1_F$ is Multiple Stochastic integrable on T^m with respect to Z. Then*

(i) *for each $y \in T^q$, except possibly on a set of quadratic variation zero with respect to Y, $f(\cdot, y)1_F(\cdot, y)$ is Multiple Stochastic integrable on T^p with respect to X; and*

(ii) *$M_m^Z(f1_F) = M_q^Y M_p^X(f1_F)$ if Y has bounded quadratic variation $V(Y)$, where the symbol $M_m^Z(f1_F)$ denotes the Multiple Stochastic integral of $f1_F$ with respect to the integrator Z on T^m.*

Proof. Given $\epsilon > 0$ there exists a positive function $\delta(x, y) > 0$ on $T^p \times T^q$ such that

$$E\left(|S(f1_F, \delta, D_1, Z) - S(f1_F, \delta, D_2, Z)|^2\right) < \epsilon$$

whenever D_1 and D_2 are standard δ-fine divisions of $T^p \times T^q$. Let N be a subset of T^q consisting of $y \in T^q$ such that $f(\cdot, y)1_F(\cdot, y)$ is not integrable on T^p. For each $y \in N$, let $\delta_y(x) = \delta_{1y}(x) = {}^{\delta(x,y)}/\sqrt{2}$ given as in Lemma 6.6. By Cauchy's criteria (Theorem 6.4), for the non-existence of integral of $f(\cdot, y)1_F(\cdot, y)$ on T^p, for each $y \in N$, there exists $Q(y) > 0$ and two standard δ_y-fine divisions of T^p, namely,

$$D_1'(y) = \{(I_y^{(i)}, x_y^{(i)})\}_{i=1}^{s(y)}$$

and

$$D_2'(y) = \{(J_y^{(i)}, u_y^{(i)})\}_{i=1}^{l(y)}$$

such that

$$\begin{aligned} 0 < Q(y) \le E\bigg(\bigg|&(D_1'(y)) \sum f(x_y^{(i)}, y)1_F(x_y^{(i)}, y)X(I_y^{(i)}) \\ &- (D_2'(y)) \sum f(u_y^{(i)}, y)1_F(u_y^{(i)}, y)X(J_y^{(i)})\bigg|^2\bigg). \end{aligned} \tag{6.1}$$

We may assume that $s(y) = l(y)$, $I_y^{(i)} = J_y^{(i)}$ for all i.

In fact the inequality (6.1) can be extended to all $y \in T^q \setminus N$ with $Q(y) = 0$ by choosing any δ_y-fine division $D_1'(y) = \{(I_y^{(i)}, x_y^{(i)})\}_{i=1}^{s(y)}$ and setting $D_2'(y) = D_1'(y)$ for all $y \in T^q \backslash N$. Next we shall prove that Q is of quadratic variation zero.

Apply Lemma 5.7 to $D_1'(y)$ and $D_2'(y)$ and let

$$\delta_2(y) = \min\left\{\frac{\delta(x^{(i)}, y)}{\sqrt{2}}, \frac{\delta(u_y^{(j)}, y)}{\sqrt{2}} : i = 1, 2, \cdots, s(y), j = 1, 2, \cdots, l(y)\right\}.$$

Denote δ_2 by δ' and let $D_j = \{(K^{(j)}, y^{(j)})\}_{j=1}^r$ be a standard δ'-fine division of T^q. By Lemma 6.6, $\left(I_{y(j)}^{(i)} \times K^{(j)}, (x_{y(j)}^{(i)}, y^{(j)})\right)$ and $\left(J_{y(j)}^{(i)} \times K^{(j)}, (u_{y(j)}^{(i)}, y^{(j)})\right)$ are δ-fine for all i and j. Since F is contained in one contiguous set, which is open, assume that $I_{y(j)}^{(i)} \times K^{(j)}$ lies completely in G_π whenever $(x_{y(j)}^{(i)}, y^{(j)}) \in F$ for the particular choice of $\delta(\xi)$ on $T^p \times T^q$.

By (6.1) and using Lemma 6.4(i), we have

$$\begin{aligned}
0 &\leq \sum_j \left[Q(y^{(j)})E(Y^2(K^{(j)}))\right] \\
&\leq \sum_j E(Y^2(K^{(j)})) \\
&\quad E\left(\left|S(f1_F, \delta_{y(j)}, D_1'(y^{(i)}), X) - S(f1_F, \delta_{y(j)}, D_2'(y^{(j)}), X)\right|^2\right) \\
&= E\left(\left|\sum_j \left(S(f1_F, \delta_{y(j)}, D_1'(y^{(j)}), X)\right.\right.\right. \\
&\quad \left.\left.\left. - S(f1_F, \delta_{y(j)}, D_2'(y^{(j)}), X)\right)Y(K^{(j)})\right|^2\right) \\
&< \epsilon,
\end{aligned} \tag{6.2}$$

thus the function Q is of quadratic variation zero with respect to Y. Hence the set N is of zero quadratic variation with respect to Y.

Next we shall prove (ii). Without loss of generality assume that the function given by $M_p^X(f(\cdot, y)1_F(\cdot, y))$ exists for all $y \in T^q$. For each $\epsilon > 0$ there exists $\delta(x, y) > 0$ on $T^p \times T^q$ such that

$$E\left((D)\left|\sum_i f(x^{(i)}, y^{(i)})1_F(x^{(i)}, y^{(i)})Z(I^{(i)}) - M_m^Z(f1_F)\right|^2\right) < \epsilon \tag{6.3}$$

whenever $D = \{(I^{(i)}, (x^{(i)}, y^{(i)}))\}_{i=1}^n$ is a standard δ-fine division of $T^p \times T^q$. For each $y \in T^q$, let δ_{1y} be as given in Lemma 6.6. We may assume that for any δ_{1y}-fine division $D_1(y) = \{(I_y^{(i)}, x_y^{(i)})\}_{i=1}^{n(y)}$ of T^p we have

$$E\left|(D_1(y))\sum f(x_y^{(i)}, y)1_F(x_y^{(i)}, y)X(I_y^{(i)}) - M_p^X(f(\cdot, y)1_F(\cdot, y))\right|^2 < \epsilon.$$

Let $\delta_2(y)$ be also as that given in Lemma 6.6. By the same lemma, the division given by $\{((I_y^{(i)} \times K^{(j)}), (x_y^{(i)}, y^{(j)}))\}_{i,j=1}^{n(y^{(j)}),r}$ is a standard δ-fine division of $T^p \times T^q$ whenever $D_2 = \{(K^{(j)}, y^{(j)})\}_{j=1}^{r}$ is a standard δ_2-fine division of T^q. Therefore

$$E\left(\left|\sum_j \sum_i f(x_{y(j)}^{(i)}, y^{(j)}) 1_F(x_{y(j)}^{(i)}, y^{(j)}) X(I_y^{(i)}) Y(K^{(j)}) - M_m^Z(f1_F)\right|^2\right) < \epsilon.$$

Applying Lemma 6.4 (ii) with $a_{ij} = f\big(x_{y(j)}^{(i)}, y^{(j)}\big) 1_F(x_{y(j)}^{(i)}, y^{(i)}\big)$, we get

$$\begin{aligned}
&E\left(\left|\sum_j \sum_i f(x_{y(j)}^{(i)}, y^{(j)}) 1_F(x_{y(j)}^{(i)}, y^{(j)}) X(I_{y(j)}^{(i)})\right.\right.\\
&\left.\left.- \sum_j M_p^X(f(\cdot, y^{(j)}) 1_F(\cdot, y^{(j)}))[Y(K^{(j)})]\right|^2\right)\\
&= \sum_j \left(E[Y^2(K^{(j)})] \right. \qquad (6.4)\\
&\left. E\left[\sum_i (f(x_{y(j)}^{(i)}, y^{(i)}) 1_F(x_{y(j)}^{(i)}) X(I_{y(j)}^{(i)}) - M_p^X(f(\cdot, y^{(j)}) 1_F(\cdot, y^{(j)}))\right]^2\right)\\
&< \epsilon V(Y),
\end{aligned}$$

where $V(Y)$ denotes the quadratic variation of Y, see Definition 6.6 above. Consequently

$$E\left|\sum_j M_p(f(\cdot, y^{(j)})) 1_F(\cdot, y^{(j)}) Y(K^{(j)}) - M_m^Z(f1_F)\right|^2 \leq 2\epsilon V(Y) + 2\epsilon$$

whenever $D_2 = \{(K^{(j)}, y^{(j)})\}_{j=1}^{r}$ is a standard δ_2-fine division of T^q. Hence

$$M_m^Z(f1_F) = M_q^Y M_p^X(f(\cdot, y) 1_F(\cdot, y))$$

thereby completing the proof. □

Theorem 6.7 (Iterated Wiener Integral). *Let G_π be an open set of T^m mentioned in Theorem 6.6 above. Suppose that $f1_{G_\pi}$ is Itô–McShane integrable on T^m with value $IM_m(f1_{G_\pi})$. Then*

$$\begin{aligned}
&IM_m(f1_{G_\pi})\\
&= \int_0^1 \int_0^{x_{\pi(m)}} \cdots \int_0^{x_{\pi(2)}} f(t_1, t_2, \ldots, t_m) 1_{G_\pi} dB_{t_{\pi(1)}} dB_{t_{\pi(2)}} \cdots dB_{t_{\pi(m)}}.
\end{aligned}$$

Proof. Applying Henstock–Fubini's Theorem (Theorem 6.6) to $Z(I \times J) = B(I \times J) = B(I)B(J)$, we get

$$
\begin{aligned}
& IM_m(f1_{G_{\pi(m)}}) \\
&= IM_{m-1}(IM_1(f(t_{\pi(1)},\cdot)1_{G_\pi}(t_{\pi(1)},\cdot)) \\
&= IM_{m-1}\left(\int_0^{x_{\pi(2)}} f(t_{\pi(1)},\cdot)1_{G_\pi}(t_{\pi(1)},\cdot)dB_{t_{\pi(1)}}\right) \\
&= IM_{m-2}\left(\int_0^{x_{\pi(3)}}\int_0^{x_{\pi(2)}} f(t_{\pi(1)},t_{\pi(2)},\cdot)1_{G_\pi}(t_{\pi(1)},t_{\pi(2)},\ldots)dB_{t_{\pi(1)}}dB_{t_{\pi(2)}}\right) \\
&\ \ \vdots \\
&= \int_0^1\int_0^{x_{\pi(m)}}\cdots\int_0^{x_{\pi(3)}}\int_0^{x_{\pi(2)}} f(t_{\pi(1)},t_{\pi(2)},\cdots,t_{\pi(m)}) \\
&\qquad 1_{G_\pi}(t_{\pi(1)},t_{\pi(2)}\cdots,t_{\pi(m)})dB_{t_{\pi(1)}}dB_{t_{\pi(2)}}\cdots dB_{t_{\pi(m)}}
\end{aligned}
$$

thereby completing the proof. □

We have used Henstock's approach to derive the Henstock–Fubini's Theorem for the Multiple Stochastic Integral, the idea of which was inspired by the classical Integration Theory approach. We further remark that Henstock's approach can also be used to study the integral over the diagonal of T^m, and the classical Hu–Meyer Theorem can be derived. This will be discussed in the next section.

6.2 Integration on the Diagonal and Hu–Meyer Theorem

Recall that in $\mathbb{R}^m$, the diagonal set consists of those points which contain some identical coordinates, that is,

$$\mathcal{D} = \{(x_1, x_2, \cdots, x_m) \in \mathbb{R}^m : x_i = x_j \text{ for some } i \neq j\}.$$

6.2.1 *Integration on the diagonal*

Remark 6.1. It is well-known from Probability Theory that if U is a normal random variable with mean 0 and variance σ^2, then $E(U^{2n+1}) = 0$ and $E(U^{2n}) = c_n\sigma^{2n}$ for any positive integer n, where $c_n = \frac{(2n)!}{2^n \cdot n!}$, see, e.g., [Oksendal (1996), Example 2.7, p. 15] and [Stromberg (1994), p. 280, Appendix to Chapter 8]. Note that c_n is increasing in n. If U is the random variable $N(0,h)$, then for the convenience of our presentation we shall write

$$E(U^p) = \langle c_{p/2}\rangle h^{p/2}$$

where $\langle c_{p/2}\rangle = c_{p/2}$ if p is even and $\langle c_{p/2}\rangle = 0$ if p is an odd integer.

Next we shall consider McShane's integral on a diagonal of $\mathbb{R}^m$.

Definition 6.9. Consider a diagonal of the form

$$\mathcal{B} = \{(\underbrace{x_1, \ldots, x_1}_{\alpha_1 terms}, \ldots, \underbrace{x_r, \ldots, x_r}_{\alpha_r terms}) \in T^m : x_i \in \mathbb{R},$$
$$i = 1, 2, \ldots, r, \; x_i \neq x_j \text{ if } i \neq j\}$$

where $\alpha_1 + \alpha_2 + \cdots + \alpha_r = m$. The *order* of the diagonal $\mathcal{B}$ is the value of

$$\max(\alpha_1, \alpha_2, \cdots, \alpha_r).$$

Without loss of generality, we assume α_1 to be the order of the diagonal.

Subsequently the integrand f that we will be considering is symmetric. Hence our consideration of the positions of the equal coordinates becomes immaterial. For the convenience of our presentation, we shall present the diagonal $\mathcal{B}$ in the above form by cluttering all the identical ordinates together as in $\mathcal{B}$.

We state and proof Lemma 6.7, which is needed for us to consider the integral on the diagonal.

Lemma 6.7. *Let* $\{\beta_1, \beta_2, \ldots, \beta_s\}$ *be a set of positive integers with* $s \leq 2r$ *and* $\beta_1 + \beta_2 + \cdots + \beta_s = 2m$. *Let*

$$\{D_i : i = 1, 2, \ldots, n\} = \{\{I_{i_1}, I_{i_2}, \ldots, I_{i_s}\} : i = 1, 2, \ldots, n\}$$

be a family of collections of subintervals of $[0, 1]$ *where* $I_{i_k} \cap I_{i_l} = \phi$ *for all* $k \neq l$ *and where* $k, l = 1, 2, \ldots, s$ *and that* $I_{i_k} \cap I_{j_l} = \phi$ *or* $I_{i_k} = I_{j_l}$ *for all* $i \neq j$, $i, j = 1, 2, \ldots, n$ *and all* $k \neq l$ *where* $k, l = 1, 2, \ldots, s$. *Let* $0 < \epsilon \leq 1$ *be given and*

$$Q = \sum_{i=1}^{n} a_{i_1} a_{i_2} E\left[B^{\beta_1}(I_{i_1}) B^{\beta_2}(I_{i_2}) \cdots B^{\beta_s}(I_{i_s})\right].$$

Then $|Q| < \langle c_{2m} \rangle^{2r} \epsilon$ *if either of the following conditions hold true:*

(i) *if there exist* k *and* l *with* $k \neq l$ *for some* $k, l = 1, 2, \ldots, s$ *such that* $\beta_k, \beta_l \geq 4$ *and that*

$$|I_{i_k}| < \frac{\sqrt{\epsilon}}{|a_{i_1}| + 1} \quad \text{and} \quad |I_{i_l}| < \frac{\sqrt{\epsilon}}{|a_{i_2}| + 1}$$

for all $i = 1, 2, \ldots, n$; *or*

(ii) *if there exists* k *such that* $\beta_k \geq 6$ *and*

$$|I_{i_k}| < \min\left\{\frac{\sqrt{\epsilon}}{|a_{i_1}| + 1}, \frac{\sqrt{\epsilon}}{|a_{i_2}| + 1}\right\}$$

for all $i = 1, 2, \ldots, n$.

Proof. It is sufficient for either case to assume that all β_i, where $i = 1, 2, \ldots, s$, to be even; otherwise $Q = 0$ and the result is immediate. We shall use the notation that if $A \in \mathbb{R}$ then $|A|$ denotes the absolute value of A and if $A \subset \mathbb{R}$, then $|A|$ is the Lebesgue measure of A; in particular, if A is an interval, then $|A|$ is the length of the interval A. We have

$$\begin{aligned} |Q| &= \left| \sum_{i=1}^{n} a_{i_1} a_{i_2} E\left[B^{\beta_1}(I_{i_1}) B^{\beta_2}(I_{i_2}) \cdots B^{\beta_s}(I_{i_s}) \right] \right| \\ &\leq \langle c_{2m} \rangle^s \sum_{i=1}^{n} |a_{i_1}||a_{i_2}||I_{i_1}|^{\frac{\beta_1}{2}} |I_{i_2}|^{\frac{\beta_2}{2}} \cdots |I_{i_s}|^{\frac{\beta_s}{2}}. \end{aligned} \tag{6.5}$$

Note that (6.5) is true by Remark 6.1 and that $\langle c_k \rangle$ is an increasing sequence in k, hence $\langle c_k \rangle \leq \langle c_{2m} \rangle$ for all $k \leq 2m$. To prove case (i), without loss of generality we may assume that $\beta_1, \beta_2 \geq 4$, so that we have ${}^{\beta_1}/_2 - 1 \geq 1$ and ${}^{\beta_2}/_2 - 1 \geq 1$ while ${}^{\beta_k}/_2 \geq 1$ for $k = 3, 4, \ldots, s$, since all these β_k's are even.

$$\begin{aligned} |Q| &\leq \langle c_{2m} \rangle^s \sum_{i=1}^{n} |a_{i_1}||a_{i_2}||I_{i_1}||I_{i_2}||I_{i_1}|^{\frac{\beta_1}{2}-1} |I_{i_2}|^{\frac{\beta_2}{2}-1} |I_{i_3}|^{\frac{\beta_3}{2}} |I_{i_4}|^{\frac{\beta_4}{2}} \cdots |I_{i_s}|^{\frac{\beta_s}{2}} \\ &< \langle c_{2m} \rangle^s \sum_{i=1}^{n} |a_{i_1}||a_{i_2}| \frac{\sqrt{\epsilon}}{|a_{i_1}| + 1} \frac{\sqrt{\epsilon}}{|a_{i_2}| + 1} |I_{i_1}||I_{i_2}| \cdots |I_{i_s}| \\ &\leq \langle c_{2m} \rangle^s \epsilon \sum_{i=1}^{n} |I_{i_1}||I_{i_2}| \cdots |I_{i_s}| \\ &\leq \epsilon \langle c_{2m} \rangle^s \left| [0,1]^s \right| \\ &\leq \langle c_{2m} \rangle^{2r} \epsilon \end{aligned}$$

thereby completing the proof for case (i).

Next to prove case (ii), assume that $\beta_1 \geq 6$ so that ${}^{\beta_1}/_2 - 2 \geq 1$. Then

$$\begin{aligned} |Q| &\leq \sum_{i=1}^{n} |a_{i_1}||a_{i_2}||I_{i_1}|^2 |I_{i_1}|^{\frac{\beta_1}{2}-2} |I_{i_2}|^{\frac{\beta_2}{2}} |I_{i_3}|^{\frac{\beta_3}{2}} \cdots |I_{i_s}|^{\frac{\beta_s}{2}} \\ &< \langle c_{2m} \rangle^s \epsilon \sum_{i=1}^{n} |I_{i_1}||I_{i_2}| \cdots |I_{i_s}| \\ &\leq \langle c_{2m} \rangle^s \left| [0,1]^s \right| \epsilon \\ &\leq \langle c_{2m} \rangle^{2r} \epsilon \end{aligned}$$

thereby completing the proof of case (ii). $\square$

Now we modify Definition 5.2 and give the following definition. Recall that in Definition 5.2, the Itô–McShane integral does not take into account the values of the integrand f on the diagonal. Now in the following definition, the Wiener–McShane integral takes care of both the diagonal and non-diagonal.

Definition 6.10. A function $f : T^m \to \mathbb{R}$ is said to be Multiple Wiener–McShane integrable to a L^2-random variable $WM_m(f)$ on T^m if for every $\epsilon > 0$, there exists a positive function δ such that

$$E\left(|S(f,\delta,D) - WM_m(f)|^2\right) < \epsilon$$

whenever $D = \{(I^{(i)}, x^{(i)})\}_{i=1}^n$ is a δ-fine division of T^m.

Note that f is Multiple Itô–McShane integrable on T^m if and only if f_0 is Multiple Wiener–McShane integrable on T^m.

Theorem 6.8. *Let $f : T^m \to \mathbb{R}$ be a real-valued function and let*

$$\mathcal{B} = \{(\underbrace{x_1,\ldots,x_1}_{\alpha_1 terms}, \underbrace{x_2,\ldots,x_2}_{\alpha_2 terms}, \ldots, \underbrace{x_r,\ldots,x_r}_{\alpha_r terms}) \in T^m : x_1 < x_2 < \cdots < x_r\}$$

be part of a diagonal of order $p \geq 3$. If the Wiener–McShane integral of $f1_{\mathcal{B}}$ exists. Then its value $WM_m(f1_{\mathcal{B}})$ equals zero.

Proof. Without loss of generality, assume $p = \alpha_1$. Given $\epsilon \in (0,1)$ there exists $\delta(\xi) > 0$ such that

$$E\left(\left|(D)\sum_{i=1}^{n} f(\xi^{(i)})1_{\mathcal{B}}(\xi^{(i)})B(I^{(i)}) - WM_m(f1_{\mathcal{B}})\right|^2\right) < \epsilon$$

for any standard δ-fine division $D = \{(I^{(i)}, \xi^{(i)})\}_{i=1}^n$ of T^m where $B(I^{(i)}) = \prod_{j=1}^m B(I_j^{(i)})$ and $I^{(i)} = \prod_{j=1}^m I_j^{(i)}$. We may assume that $\delta(\xi) > 0$ is chosen such that any δ-fine interval with associated point $\xi \in G_\pi$ for some $\pi \in G_\pi$ lies completely in G_π. This is possible since each G_π is open.

We consider a special division of T^m such that if $\xi \in \mathcal{B}$, then the associated interval is of the form $\prod_{j=1}^r I_{i_j}^{\alpha_j}$. Let the corresponding δ-fine partial division (with the corresponding points $\xi \in \mathcal{B}$) be

$$P = \left\{\left(\prod_{j=1}^{r} I_{i_j}^{\alpha_j}, \xi^i\right) : \xi^i \in \mathcal{B}, i = 1,2,\ldots,n\right\}.$$

If we can show that

$$Q = E\left(\left|(P)\sum_{i=1}^{n} f(\xi^i)1_{\mathcal{B}}(\xi^i)\prod_{j=1}^{r} B^{\alpha_j}(I_{i_j})\right|^2\right) < K\epsilon$$

for some positive real value K, then we would have shown that $WM_m(f1_{\mathcal{B}}) = 0$.

$$\begin{aligned} Q &= E\left(\left|(P)\sum_{i=1}^{n} f(\xi^i)1_{\mathcal{B}}(\xi^i)\prod_{j=1}^{r} B^{\alpha_j}(I_{i_j})\right|^2\right) \\ &= E\left\{(P)\sum_{i=1}^{n} f^2(\xi^i)\prod_{j=1}^{r} B^{2\alpha_j}(I_{i_j})\right\} \\ &\quad + 2E\left\{(P)\sum_{i<j} f(\xi^i)f(\xi^j)\prod_{l=1}^{r} B^{\alpha_l}(I_{i_l})\prod_{l=1}^{r} B^{\alpha_l}(I_{j_l})\right\} \\ &= Q_1 + Q_2, \end{aligned}$$

where

$$|Q_1| = \sum_{i=1}^{n} f^2(\xi^i)E\left[\prod_{j=1}^{r} B^{2\alpha_j}(I_{i_j})\right].$$

Now that $\alpha_1 \geq 3$ implies that $2\alpha_1 \geq 6$. By case (ii) of Lemma 6.7 take $a_{i_1} = a_{i_2} = f(\xi^i)$ and $\delta(\xi^i) \leq \sqrt{\epsilon}/(|f(\xi^i)|+1)$ so that we have $|Q_1| \leq \langle c_{2m}\rangle^{2r}\epsilon$.

Consider

$$\begin{aligned} Q_2 &= \sum_{i<j} f(\xi^i)f(\xi^j)E\left[\prod_{l=1}^{r} B^{\alpha_l}(I_{i_l})\prod_{l=1}^{r} B^{\alpha_l}B(I_{j_l})\right] \\ &= \sum_{i<j} f(\xi^i)f(\xi^j)E\left[\prod_{k=1}^{s} B^{\beta_k}(I_{u_k})\right], \end{aligned}$$

where $\beta_1 + \beta_2 + \cdots + \beta_s = 2m$. We only need to consider those pairs of integers $\{i, j\}$ such that all $\beta_k, k = 1, 2, \ldots, s$ are even or otherwise $Q_2 = 0$. We may also assume that $\beta_1, \beta_2 \geq 4$ since $\alpha_1 \geq 3$, so that choose $I_{u_1} = I_{i_l}$ for some l and $I_{u_2} = I_{j_k}$ for some k, which are needed when applying Lemma 6.7 (see the last two inequalities of this proof). Consequently we may write

$$Q_2 = \sum_{s=r+1}^{2r} \sum_{\{\beta_1,\beta_2,\ldots,\beta_s\}} \left(\sum_{i<j} f(\xi^i)f(\xi^j)E\left[\prod_{k=1}^{s} B^{\beta_k}(I_{u_k})\right]\right),$$

where we remark that (i) for each $s = r+1, r+2, \ldots, 2r$ the number of even integer solutions for the equation

$$\beta_1 + \beta_2 + \cdots + \beta_s = 2m$$

does not exceed $\binom{m-1}{s-1}$, and (ii) for each product

$$B^{\beta_1}(I_{u_1})B^{\beta_2}(I_{u_2}) \cdots B^{\beta_s}(I_{u_s})$$

can be the result of the product of two intervals of the form $I_{u_1}^{\alpha_1} \times I_{u_2}^{\alpha_2} \times \cdots \times I_{u_s}^{\alpha_s}$ and $I_{u_1}^{\beta_1-\alpha_1} \times I_{u_2}^{\beta_2-\alpha_2} \times \cdots \times I_{u_s}^{\beta_s-\alpha_s}$ where $\alpha_1 + \alpha_2 + \cdots + \alpha_s = m$ and each α_i is a non-negative integer. The number of such possible choices of $\{\alpha_1, \alpha_2, \ldots, \alpha_s\}$ cannot exceed $\binom{m+s-1}{s-1}$ by simple combinatorics. Consequently

$$|Q_2| < \sum_{s=r+1}^{2r} \binom{m-1}{s-1}\binom{m+s-1}{s-1} \sum f(\xi^i)f(\xi^j)E\left(\prod_{k=1}^{s} B^{\beta_k}(I_{u_k})\right)$$

where $\sum f(\xi^i)f(\xi^j)E\left(\prod_{k=1}^{s} B^{\beta_k}(I_{u_k})\right)$ denote the summation over all possible $I_{u_1} \times I_{u_2} \times \cdots \times I_{u_s}$ for each fixed set $\{\beta_1, \beta_2, \ldots, \beta_s\}$. We may assume that $\delta(\xi^i) \leq \sqrt{\epsilon}/(|f(\xi^i)|+1)$ so that by Lemma 6.7(i) we have

$$|Q_2| < \langle c_{2m} \rangle^{2r} \left(\sum_{s=r+1}^{2r} \binom{m-1}{s-1}\binom{m+s-1}{s-1}\right) \epsilon$$

thereby completing our proof. □

Remark 6.2. Let $f : T^m \to \mathbb{R}$ be a symmetric function. From Theorem 6.8, it suffices to consider only diagonals in T^m of order 2 whenever we want to investigate the integral of f on diagonals of T^m.

For each $j = 1, 2, \ldots, \left[\frac{m}{2}\right]$, let

$$\mathcal{B}_j = \{(\underbrace{x_1, x_1}_{y_1}, \underbrace{x_2, x_2}_{y_2}, \ldots, \underbrace{x_j, x_j}_{y_j}, \underbrace{x_{2j+1}}_{y_{j+1}}, \underbrace{x_{2j+2}}_{y_{j+2}}, \ldots, \underbrace{x_m}_{y_{m-j}}) \in T^m :$$

$$x_i \in [0,1]; x_i \neq x_k \text{ if } i \neq k\}$$

denote a diagonal of order 2 with exactly j pairs of equal components.

6.2.2 *Traces and Hu–Meyer Theorem*

Let $f_p : T^{m-j} \to \mathbb{R}$ denote the nondiagonal projection of the function f, that is,

$$f_p(y_1, y_2, \ldots, y_{m-j}) = f(x_1, x_1, \ldots, x_j, x_j, x_{2j+1}, \ldots, x_m)$$

for all $(x_1, x_1, x_2, x_2, \ldots, x_j, x_j, x_{2j+1}, \ldots, x_m) \in \mathcal{B}_j$ and where $y_1 = x_1$, $y_2 = x_2, \ldots, y_j = x_j, y_{j+1} = x_{2j+1}, \ldots y_{m-j} = x_m$.

Suppose that the function $f_p(\cdot, y_{j+1}, y_{j+2}, \ldots, y_{m-j})$ is Lebesgue integrable on T^j for each point $(y_{j+1}, y_{j+2}, \ldots, y_{m-j}) \in T^{m-2j}$. Define the function $\mathrm{tr}_j\{f\} : T^{m-2j} \to \mathbb{R}$ as

$$\begin{aligned}&\mathrm{tr}_j\{f\}(y_{j+1}, \ldots, y_{m-j})\\&= (L)\int_{T^j} f_p(y_1, y_2, \ldots, y_j, y_{j+1}, \ldots, y_{m-j})dy_1 dy_2 \cdots dy_j,\end{aligned}$$

where $(L)\int$ denotes the Lebesgue integral of f_p with respect to the first j components of f_p. We shall call the function $\mathrm{tr}_j\{f\}$ the *trace* of f of order j, which is slightly different from the classical definition, see [Nualart (1995); Sole and Utzet (1990)]. Recall that since all integrands that we are considering is symmetrical, the positions of the coordinates of the j pairs of equal coordinates is immaterial.

We next need to prove the following lemma which is an important property that is used in the proof of Theorem 6.9.

Lemma 6.8. *Let $J = \prod_{s=1}^{j} J_s^2 \subset T^{2j}$, where each J_s is an interval from $T = [0, 1]$ and $\{J_s\}$ are non-overlapping. Let $\lambda_p(J)$ denote $\prod_{s=1}^{j} |J_s|$. Then*

(i) *$E[B(J)] = \lambda_p(J)$; and*
(ii) *$E[B^2(J)] = 3^j [\lambda_p(J)]^2$.*

Proof. For each interval $J_s \subset \mathbb{R}$ we have $E[B^2(J_s)] = |J_s|$ and $E[B^4(J_s)] = 3|J_s|^2$, see Remark 6.1, and that B has orthogonal increments. Hence part (i) follows since

$$E[B(J)] = E\left[\prod_{s=1}^{j} B(J_s^2)\right] = \prod_{s=1}^{j} E[B^2(J_s)] = \lambda_p(J)$$

and (ii) follows from that

$$E[B^2(J)] = E\left[\prod_{s=1}^{j} B^2(J_s^2)\right] = E\left[\prod_{s=1}^{j} B^4(J_s^2)\right] = 3^j [\lambda_p(J)]^2$$

thereby completing our proof. □

Theorem 6.9. *Let $\mathcal{B}_j$ be a diagonal of order 2 and with exactly j pairs of equal components, that is,*

$$\mathcal{B}_j = \{(\underbrace{x_1, x_1}_{y_1}, \underbrace{x_2, x_2}_{y_2}, \dots, \underbrace{x_j, x_j}_{y_j}, \underbrace{x_{2j+1}}_{y_{j+1}}, \underbrace{x_{2j+2}}_{y_{j+2}}, \dots, \underbrace{x_m}_{y_{m-j}}) \in T^m : x_1 < x_2 < \cdots < x_m\}$$

and $f : T^m \to \mathbb{R}$ be a symmetric real-valued function. Suppose that for each point $(y_{j+1}, y_{j+2}, \dots, y_{m-j}) \in T^{m-2j}$, the trace function $\mathrm{tr}_j\{f\} : T^{m-2j} \to \mathbb{R}$ exists, that is, $f_p : T^{m-j} \to \mathbb{R}$ is Lebesgue integrable on T^j for each point $(y_{j+1}, y_{j+2}, \dots, y_{m-j}) \in T^{m-2j}$, and that $f_p(\cdot, y_{j+1}, y_{j+2}, \cdots, y_{m-j})$ is square Lebesgue integrable for each point $(y_{j+1}, y_{j+2}, \dots, y_{m-j}) \in T^{m-2j}$. Further suppose that $f1_{\mathcal{B}_j}$ is Multiple Wiener–McShane integrable on T^m and that $\mathrm{tr}_j\{f\}$ is Multiple Wiener–McShane integrable on T^{m-2j}. Then

$$WM_m\left[f1_{\mathcal{B}_j}\right] = WM_{m-2j}\left[\mathrm{tr}_j\{f\}\right].$$

Proof. Given $\epsilon > 0$ there exists $\delta_1(x) > 0$ on T^m such that

$$E\left(\left|(D_1)\sum_{i=1}^{n} f(x^{(i)})B(J^{(i)} \times I^{(i)}) - WM_m(f1_{\mathcal{B}_j})\right|^2\right) < \epsilon$$

for any standard δ_1-fine division $D_1 = \{(J^{(i)} \times I^{(i)}, x^{(i)})\}_{i=1}^{n}$ of T^m, where $J^{(i)} \subset T^{2j}$ and $I^{(i)} \subset T^{m-2j}$ for each $i = 1, 2, \dots, n$.

For each fixed $y \in T^{m-2j}$, choose $\delta_2(\tilde{y}, y) > 0$ for each $\tilde{y} \in T^j$ such that

$$\left|(D_2)\sum f_p(y^{(i)}, y)|J^{(i)}| - \mathrm{tr}_j\{f\}(y)\right| < \sqrt{\epsilon} \tag{6.6}$$

for any δ_2-fine division of T^j denoted by

$$D_2(y) = \{(J^{(i)}, y^{(i)})\}_{i=1}^{n(y)}$$

so that δ_2 is a function defined on T^j depending on each $y \in T^{m-2j}$. Since δ_2 is a function defined on T^j for each $y \in T^{m-2j}$, we can see that δ_2 is a function on T^{m-j}.

Choose a function δ on T^m such that $\delta \le \delta_1$ and that $\delta|_{\mathcal{B}_j} \le \delta_2$, that is, for any

$$x = (\underbrace{x_1, x_1}, \underbrace{x_2, x_2}, \dots, \underbrace{x_j, x_j}, x_{2j+1}, x_{2j+2}, \dots, x_m) \in \mathcal{B}_j,$$

we have $\delta(x) \le \delta_2(x_1, x_2, \dots, x_j, x_{2j+1}, \dots, x_m)$.

Since $\operatorname{tr}\{f\}$ is Multiple Wiener–McShane integrable on T^{m-2j}, let $\delta_3(y) > 0$ be defined on T^{m-2j} such that

$$E\left(\left|(D_3)\sum_{k=1}^{q}\operatorname{tr}_j\{f\}(y^{(k)})B(I^{(k)}) - WM_{m-2j}(\operatorname{tr}_j\{f\})\right|^2\right) < \epsilon$$

for any standard $\delta_3(y)$-fine division $D_3 = \{(I^{(k)}, y^{(k)})\}_{k=1}^{q}$ of T^{m-2j}.

Recall that $T^m = T^{2j} \times T^{m-2j}$. For each $y \in T^{m-2j}$, define $\delta_{1y}(x) = \delta(x,y)/\sqrt{2}$ for all $x \in T^{2j}$. Let $D_1(y) = \{(J_y^{(i)}, x_y^{(i)})\}_{i=1}^{m(y)}$ be a δ_{1y}-fine division of T^{2j}. Then, by Lemma 6.6, there exists $\delta'(y) > 0$ on all $y \in T^{m-2j}$ such that $(J_y^{(i)} \times I, (x^{(i)}, y))$ is δ-fine if (I, y) is δ'-fine on T^{m-2j}. We may choose $\delta' \leq \delta_3$ on T^{m-2j}.

Let $D_p = \{(I_k, y_k)\}_{k=1}^{q}$ be a standard δ'-fine division on T^{m-2j} such that

$$D_a = \{(J_{y_k}^{(i)} \times I_k, (x_{y_k}^{(i)}, y_k))\}_{i,k=1}^{n(y_k),q}$$

is standard δ-fine of T^m. Choose such a division D_a such that each $J_{y_k}^{(i)}$ is of the form $\prod_{s=1}^{j} J_{k_s}^{(i)} \times J_{k_s}^{(i)}$ and that $x_{y_k}^{(i)} = (x_1, x_1, x_2, x_2, \ldots, x_j, x_j) \in T^{2j}$ such that $x_{p(y_k)}^{(i)} = (x_1, x_2, \ldots, x_j) \in T^j$, hence

$$f(x_{y_k}^{(i)}, y_k) = f_p(x_{p(y_k)}^{(i)}, y_k).$$

Consequently

$$\begin{aligned}
&E|WM_m(f1_{\mathcal{B}_j}) - WM_{m-2j}(\operatorname{tr}_j\{f\})|^2 \\
&\leq 3\left|WM_m(f1_{\mathcal{B}_j}) - (D_a)\sum_{k=1}^{q}\sum_{i=1}^{n(y_k)} f(x_{y_k}^{(i)}, y_k)1_{\mathcal{B}_j}(x_{y_k}^{(i)}, y_k)B(J_{y_k}^{(i)} \times I_k)\right|^2 \\
&\quad + 3E\left|(D_a)\sum_{k=1}^{q}\sum_{i=1}^{n(y_k)} f(x_{y_k}^{(i)}, y_k)1_{\mathcal{B}_j}(x_{y_k}^{(i)}, y_k)B(J_{y_k}^{(i)} \times I_k)\right. \\
&\qquad \left. - (D_p)\sum_{k=1}^{q}\operatorname{tr}_j\{f\}(y_k)B(I_k)\right|^2 \qquad (6.7) \\
&\quad + 3E\left|(D_p)\sum_{k=1}^{q}\operatorname{tr}\{f\}(y_k)B(I_k) - WM_{m-2j}(\operatorname{tr}_j\{f\})\right|^2 \\
&< 3\epsilon + 3Q + 3\epsilon.
\end{aligned}$$

We are using "special" division, each $J_{y_k}^{(i)}$ can be written as $\prod_{s=1}^{j}\left(J_{k_s}^{(i)} \times J_{k_s}^{(i)}\right)$, so that

$$Q = E\left|\sum_{k=1}^{q}\left\{\sum_{i=1}^{n(y_k)} f(x_{y_k}^{(i)}, y_k)1_{\mathcal{B}_j}(x_{y_k}^{(i)}, y_k)B(J_{y_k}^{(i)} \times I_k) - \text{tr}\{f\}(y_k)B(I_k)\right\}\right|^2 \tag{6.8}$$

$$= E\left|\sum_{k=1}^{q} B^2(I_k)\left\{\sum_{i=1}^{n(y_k)} f(x_{y_k}^{(i)}, y_k)1_{\mathcal{B}_j}(x_{y_k}^{(i)}, y_k)B(J_{y_k}^{(i)}) - \text{tr}\{f\}(y_k)\right\}\right|^2 \tag{6.9}$$

$$= E\left[\sum_{k=1}^{q} B^2(I_k)\left\{\sum_{i=1}^{n(y_k)} f(x_{y_k}^{(i)}, y_k)B(J_{y_k}^{(i)}) - \text{tr}\{f\}(y_k)\right\}^2\right] \tag{6.10}$$

$$= \sum_{k=1}^{q} E[B^2(I_k)]E\left\{\sum_{i=1}^{n(y_k)} f(x_{y_k}^{(i)}, y_k)1_{\mathcal{B}_j}(x_{y_k}^{(i)}, y_k)B(J_{y_k}^{(i)}) - \text{tr}\{f\}(y_k)\right\}^2. \tag{6.11}$$

Here we note that the transition from (6.9) to (6.10) is due to that

$$E\left[B(I_k)B(I_l)B(J_{y_k}^{(i)})B(J_{y_l}^{(i)})\right] = 0$$

if $J_{y_k}^{(i)} \times I_k$ and $J_{y_l}^{(i)} \times I_l$ are distinct intervals from D_a with $k \neq l$, and that (6.11) is true since $B(I_k)$ and $B(J_{y_k}^{(i)})$ are independent for each k and each $i = 1, 2, \ldots, n(y_k)$. We remark that each $J_{y_k}^{(i)}$ is of the form $J_{y_k}^{(i)} = \prod_{s=1}^{j}\left(J_{k_s}^{(i)} \times J_{k_s}^{(i)}\right)$ since we are considering standard division. Using the notations and results of Lemma 6.8,

$$\begin{aligned}
Q &\leq 2\sum_{k=1}^{q} E[B^2(I_k)]E\left\{\sum_{i=1}^{n(y_k)} f(x_{y_k}^{(i)}, y_k)1_{\mathcal{B}_j}(x_{y_k}^{(i)}, y_k)\left(B(J_{y_k}^{(i)}) - \lambda_p(J_{y_k}^{(i)})\right)\right\}^2 \\
&\quad + 2\sum_{k=1}^{q} E[^2(I_k)]E\left\{\sum_{i=1}^{n(y_k)} f(x_{y_k}^{(i)}, y_k)\lambda_p\left(J_{y_k}^{(i)}\right) - \text{tr}\{f\}(y_k)\right\}^2 \\
&< 2\sum_{k=1}^{q} E[B^2(I_k)]S(y_k) + 2\sum_{k=1}^{q} E[E^2(I_k)]\epsilon,
\end{aligned}$$

where $S(y_k)$ denotes the corresponding term in the first sum.

If we assume also that

$$\delta_{1y_k}(\underbrace{x_1,x_1},\underbrace{x_2,x_2},\dots,\underbrace{x_j,x_j}) \le \delta_2(x_1,x_2,\dots,x_j,y_k),$$

see (6.6). Now considering

$$\begin{aligned}
&\sum_{k=1}^{q} E[B^2(I_k)]S(y_k)\\
&=\sum_{k=1}^{q} E[B^2(I_k)]E\left\{\sum_{i=1}^{n(y_k)} f(x^{(i)}_{y_k},y_k)\left(B(J^{(i)}_{y_k})-\lambda_p(J^{(i)}_{y_k})\right)\right\}^2\\
&=\sum_{k=1}^{q} E[B^2(I_k)]E\left\{\sum_{i=1}^{n(y_k)} f^2(x^{(i)}_{y_k},y_k)1_{\mathcal{B}_j}(x^{(i)}_{y_k},y_k)\left(B(J^{(i)}_{y_k})-\lambda_p(J^{(i)}_{y_k})\right)^2\right\}+0\\
&\le 2\sum_{k=1}^{q} E[B^2(I_k)]\sum_{i=1}^{n(y_k)} f^2(x^{(i)}_{y_k},y_k)1_{\mathcal{B}_j}(x^{(i)}_{y_k},y_k)E[B^2(J^{(i)}_{y_k})]\\
&\quad+2\sum_{k=1}^{q} E[B^2(I_k)]\sum_{i=1}^{n(y_k)} f^2(x^{(i)}_{y_k},y_k)1_{\mathcal{B}_j}(x^{(i)}_{y_k},y_k)[\lambda_p(J^{(i)}_{y_k})]^2\\
&=2(3^j+1)\sum_{k=1}^{q} E[B^2(I_k)]\sum_{i=1}^{n(y_k)} f^2(x^{(i)}_{y_k},y_k)[\lambda_p(J_{y_k})]^2\\
&=2(3^j+1)U(y_k),
\end{aligned}$$

where $U(y_k)=\sum_{k=1}^{q} E[B^2(I_k)]\sum_{i=1}^{n(y_k)} f^2(x^{(i)}_{y_k},y_k)[\lambda_p(J_{y_k})]^2$.

Since f_p^2 is integrable on T^j, with integral denoted by $\int_{T^j} f_p^2(y)d\lambda$ for each $y\in T^{m-2j}$, we let $\delta_2(x,y)>0$ also be a function on all $x\in T^j$ associated with

$$\left|(D_4)\sum_{i=1}^{r} f_p^2(x_i,y)\prod_{s=1}^{j}|J_s^i| - \int_{T^j} f_p^2(y)d\lambda\right| < \epsilon$$

for any $\delta_2(y)$-fine division $D_4(y)=\left\{(\prod_{s=1}^{j} J_s^i, x_i)\right\}_{i=1}^{r}$ of T^j.

Choose $\delta(x,y)\le \sqrt{2}\epsilon/(\int_{T^j} f_p^2 d\lambda+1)$ so that the function δ_{1y}, by Lemma 6.6, will also satisfy $\delta_{1y}(x)\le \epsilon/(\int_{T^j} f_p^2 d\lambda+1)$ for each value of y. Hence we have that

$$\begin{aligned}U(y_k) &\le \sum_{k=1}^{q} E[B^2(I_k)] \sum_{i=1}^{n(y_k)} \lambda_p(J_{y_k}^{(i)}) f^2(x_{y_k}^{(i)}, y_k) 1_{\mathcal{B}_j}(x_{y_k}^{(i)}, y_k) \lambda_p(J_{y_k}^{(i)}) \\ &\le \sum_{k=1}^{q} E[B^2(I_k)] \frac{\epsilon}{\int_{T^j} f_p^2 d\lambda + 1} \left(\int_{T^j} f_p^2 d\lambda + \epsilon \right) \\ &< \epsilon \sum_{k=1}^{q} E[W^2(I_k)] \\ &\le \epsilon,\end{aligned}$$

thereby completing our proof.

Finally we remark that for the case when m is even and that $j = m/2$, the proof of Theorem 6.9 is much simpler and needs to start from step (6.7) above. □

We are now ready to prove the classical Hu–Meyer Theorem. However, the conditions imposed on the integrand f are slightly different from the classical Hu–Meyer Theorem, see [Nualart (1995); Sole and Utzet (1990)], since the ways to define the corresponding integrals are different.

Theorem 6.10 (Hu–Meyer Theorem). *Let $f \in L^2([0,1]^m)$ be a symmetric function. Suppose that f_0 and $f 1_{\mathcal{B}_j}$ are Multiple Wiener–McShane integrable and f has a trace of order j for each $j = 1, 2, \ldots, [m/2]$, denoted by $\mathrm{tr}_j\{f\}$, and which satisfies the conditions of Theorem 6.9. Then f is Multiple Wiener–McShane integrable and*

$$WM_m(f) = WM_m(f_0) + \sum_{j=1}^{[m/2]} \frac{m!}{(m-2j)!j!2^j} WM_{m-2j}(\mathrm{tr}_j\{f\}).$$

Proof. We partition $[0,1]^m$ into the diagonal set

$$\mathcal{D} = \{(x_1, x_2, \ldots, x_m) \in T^m : x_i = x_j \text{ for some } i \neq j\}$$

and the non-diagonal set

$$\mathcal{D}^c = \{(x_1, x_2, \ldots, x_m) \in T^m : x_i \neq x_j \text{ for all } i \neq j\}.$$

From Theorem 6.8, it is sufficient to consider the integrals over those diagonals of order 2 with j pairs of equal components, which we denote by $\mathcal{B}_j$ for each $j = 1, 2, \ldots, [m/2]$. From simple combinatorics, the number of distinct diagonals of order 2 (distinct in the sense of the difference in the positions of the j pairs of equal ordinates) and with j pairs of equal components, $j = 1, 2, \ldots, [m/2]$, can be computed as

$$\frac{1}{j!}\binom{m}{2}\binom{m-2}{2}\binom{m-4}{2}\cdots\binom{m-(2j-2)}{2} = \frac{m!}{2^j j!(m-2j)!}.$$

Since f is symmetric, its integral over each diagonal set $\mathcal{B}_j$ will be identical. Hence

$$f = f_0 + \sum_{j=1}^{[m/2]} \frac{m!}{2^j j!(m-2j)!} f 1_{\mathcal{B}_j}$$

$$\begin{aligned} WM_m(f) &= WM_m(f_0) + \sum_{j=1}^{[m/2]} \frac{m!}{2^j j!(m-2j)!} WM_m(f 1_{\mathcal{B}_j}) \\ &= WM_m(f_0) + \sum_{j=1}^{[m/2]} \frac{m!}{2^j j!(m-2j)!} WM_{m-2j}(\mathrm{tr}_j\{f\}), \end{aligned}$$

thereby completing our proof. □

6.3 Notes and Remarks

The proofs of Fubini's Theorem and Hu–Meyer's Theorem using the Henstock approach can be found in [Toh and Chew (2004–05, 2005)].

Bibliography

Bonotto, E. M., Federson, M. and Muldowney, P. (2021). *The Black–Scholes equation with impulses at random times via generalized Riemann integral*, *Proc. Singap. National Acad. Sci.*, **15**, 1, pp. 45–59. Special issue: Henstock Integration: Theory and Applications (World Scientific, Singapore).

Boonpogkrong, V. (2004). *Generalized Itô Integral and Henstock–Young Integral*, M.Sc. Thesis, NUS.

Boonpogkrong, V. (2007). *The Henstock–Kurzweil Integral with Integrators of Unbounded Variation*, Ph.D. Thesis, NUS.

Boonpogkrong, V. and Chew, T. S. (2004–05). *On Integrals with Integrators in BV_p*, *Real Anal. Exch.*, **30**, 1, pp. 193–200.

Boonpogkrong, V. and Chew, T. S. (2004). *On Nonadapted Stochastic Integrals*, *Thai J. Math.*, **2**, 1, pp. 29–39.

Boonpogkrong, V., Chew, T. S. and Toh, T. L. (Under Review). *Differentials for the Itô integral*, *Rend. Circ. Mat. Palermo*, Under Review.

Chew, T. S., Tay, J. Y. and Toh, T. L. (2001–02), *The Non-uniform Riemann approach to Ito's integral*, *Real Anal. Exch.* **27**(2), 495–514.

Canton, R. G., Labendia, M. A. and Toh, T. L. (2022). *Stratonovich–Henstock integral for the operator-value stochastic process*, *Proyecciones (Antofagasta)*, **5**, pp. 1111–1130.

Calin, O. (2015). *An Informal Introduction To Stochastic Calculus With Applications*, (World Scientific, Singapore).

Chew, T. S., Huang, Z. Y. and Wang, C. S. (2004). *The Non-uniform Riemann approach to Anticipating Stochastic Integrals*, *Stoch. Anal. Appl.*, **22**, 2, pp. 429–442.

Chew, T. S., Tay, J. Y. and Toh, T. L. (2001–02). *The Non-uniform Riemann approach to Ito's integral*, *Real Anal. Exch.*, **27**, 2, pp. 495–514.

Chung, K. L. and Williams, R. J. (1990). *Introduction to Stochastic Integration*, 2nd edn. (Birkhäuser Boston).

Dudley, R. M. and Norvaiša, R. (1999). *Differentiability of Six Operators on Nonsmooth Functions and p-Variation* (Springer).

Durrett, R. (1990). *Brownian Motion and Martingales in Analysis* (Wardsworth Advanced Books & Software, California).

Henstock, R. (1955). *The efficiency of convergence factors for functions of a continuous real variable, J. London Math. Soc.*, **30**, pp. 273–286.
Henstock, R. (1988). *Lectures on the theory of integration* (World Scientific, Singapore).
Henstock, R. (1991). *The general theory of integration* (Oxford University Press, Oxford).
Henstock, R. (1990–91). *Stochastic and other functional integral, Real Anal. Exch.*, **16**, pp. 460–470.
Ikeda, N. and Watanabe, S. (1981). *Stochastic Differential Equations and Diffusion Process* (North-Holland Mathematical Library, **24**).
Itô, K. (1944). *Stochastic integral, Proc. Imp. Acad. Tokyo*, **20**, 8, pp. 519–524.
Itô, K. (1951). *Multiple Wiener integral, J. Math. Soc. Japan*, **3**, pp. 157–169.
Klebaner, F. C. (2012). *Introduction to Stochastic Calculus with Applications*, 3rd edn. (Imperial College Press).
Kopp, P. E. (1985). *Martingales and Stochastic Integrals* (Cambridge University Press).
Kurzweil, J. (1957). *Generalised ordinary differential equations and continuous dependence on a parameter, Czech. Math. J.*, **7**, pp. 418–446.
Kurzweil, J. (2002). *Integration Between the Lebesgue Integral and the Henstock–Kurzweil Integral: Its Relation to Local Convex Vector Spaces* (World Scientific, Singapore).
Ladde, A. G. and Ladde, G. S. (2013). *An Introduction to Differential Equations: Stochastic Modeling, Methods And Analysis, Volume 2* (World Scientific).
Lee, P. Y. (1989). *Lanzhou lectures on Henstock integration* (World Scientific, Singapore).
Lee, P. Y. and Vyborny, R. (2000). *The integration: An easy approach after Kurzweil and Henstock* (Cambridge University Press).
Lee, T. Y. (2011). *Henstock–Kurzweil Integration on Euclidean Spaces* (World Scientific, Singapore).
Lim, Y. Y. C. and Toh, T. L. (2022). *A note on Henstock–Itô's non-stochastic integral, Real Anal. Exch.*, **47**, 2, pp. 1–18.
Love, E. R. and Young, L. C. (1938). *On Fractional Integration by Parts, Proc. London Math. Soc.*, **44**, pp. 1–35.
Ma, Z. M., Lee, P. Y. and Chew, T. S. (1992–93). *Absolute integration using Vitali covers, Real Anal. Exch.*, **18**, pp. 409–419.
McShane, E. J. (1969). *A Riemann-Type Integral that Includes Lebesgue–Stieltjes, Bochner and Stochastic Integrals*, (Memoirs of the American Mathematical Society **88** (Providence).
McShane, E. J. (1974). *Stochastic Calculus and Stochastic Models* (Academic Press, New York).
McShane, E. J. (1984). *Unified Integration* (Academic Press, New York).
Mikosch, T. (1998). *Elementary Stochastic Calculus with Finance in View* (World Scientific, Singapore).
Monteiro, G. A., Slavik, A. and Tvrdy, M. (2019). *Kurzweil–Stieltjes Integral: Theory And Applications* (World Scientific, Singapore).

Muldowney, P. (2012). *A Modern Theory of Random Variation: With Applications in Stochastic Calculus, Financial Mathematics, and Feynman Integration* (John Wiley & Sons).

Norin, N. V. (1996). *The Extended Stochastic Integral in Linear Spaces with Differential Measures and related topics* (World Scientific, Singapore).

Nualart, D. (1995). *The Malliavin Calculus and related topics* (Springer-Verlag, New York).

Oksendal, B. (1996). *Stochastic Differential Equations: An Introduction with Applications*, 4th edn. (Springer-Verlag).

Pop-Stojanovic, Z. R. (1972). *On McShane's belated stochastic integral*, *SIAM J. Appl. Math.*, **22**, pp. 87–92.

Protter, P. (1979). *A comparison of stochastic integrals*, *Ann. Probab.*, **7**, pp. 276–289.

Protter, P. (2005). *Stochastic integration and Differential Equations, 2nd Edition* (Springer, New York).

Revuz, D. and Yor, M. (1994). *Continuous martingales and Brownian motion*, 2nd edn. (Springer-Verlag).

Royden, H. L. (1989). *Real Analysis*, 3rd edn. (MacMillan Publishing Company, New York).

Sole, J. LL. and Utzet, F. (1990). *Stratonovich integral and trace*, *Stochastics and Stochastic Reports*, **29**, pp. 203–220.

Steele, J. M. (2001). *Stochastic Calculus and Financial Applications*, (Springer, New York).

Stromberg, K. R. (1994). *Probability for Analysts* (Chapman & Hall, New York).

Tan, S. B. and Toh, T. L. (2011–12). *The Itô–Henstock Stochastic Differential Equations*, *Real Anal. Exch.*, **37**, 2, pp. 411–424.

Toh, T. L. (2001). *The Riemann Approach to Stochastic Integration*, Ph.D. Thesis, NUS.

Toh, T. L. and Chew, T. S. (1999). *A variational approach to Ito's integral*, *Proceedings of SAP's 98*, pp. 291–299 (World Scientific, Singapore).

Toh, T. L. and Chew, T. S. (2002). *The Riemann approach to stochastic integration using non-uniform meshes*, *J. Math. Anal. Appl.*, **280**, pp. 133–147.

Toh, T. L. and Chew, T. S. (2003–04). *The non-uniform Riemann approach to the multiple Itô–Wiener integral*, *Real Anal. Exch.*, **29**, 1, pp. 275–289.

Toh, T. L. and Chew, T. S. (2005). *On belated differentiation and a characterization of Henstock–Kurzweil–Ito integrable processes*, *Math. Bohem.*, **130**, pp. 63–73.

Toh, T. L. and Chew, T. S. (2004–05). *On Henstock–Fubini's theorem for multiple Stochastic integral*, *Real Anal. Exch.*, **30**, 1, pp. 295–310.

Toh, T. L. and Chew, T. S. (2005). *On Henstock's multiple Wiener integral and Henstock's version of Hu–Mayer theorem*, *Math. Comput. Model.*, **42**, pp. 139–149.

Toh, T. L. and Chew, T. S. (2005). *On Ito–Kurzweil–Henstock integral and integration-by-part formula*, *Czech. Math. J.*, **55**, 103, pp. 656–663.

Toh, T. L. and Chew, T. S. (2010). *Henstock's version of Itô formula*, *Real Anal. Exch.*, **35**, 2, pp. 1–15.

Toh, T. L. and Chew, T. S. (2012). *The Kurzweil–Henstock theory of Stochastic integral*, *Czech. Math. J.*, **32**, pp. 1–20.

Weizsäcker, H. and Winkler, G. (1990). *Stochastic Integrals: An introduction* (Friedr Vieweg & Sohn).

Xu, J. G. and Lee, P. Y. (1992–93). *Stochastic integrals of Itô and Henstock*, *Real Anal. Exch.*, **18**, pp. 352–366.

Yang, H. and Toh, T. L. (2014). *On Henstock method to Stratonovich integral with respect to continuous semimartingale*, *Int. J. Stoch. Anal.*, pp. 1–7. (ID534864)

Yang, H. and Toh, T. L. (2016). *On Henstock–Kurzweil method to Stratonovich integral*, *Math. Bohem.*, **141**, 2, pp. 129–142.

Yeh, J. (1995). *Martingales and Stochastic Analysis* (World Scientific, Singapore).

Young, L. C. (1936). *An inequality of the Hölder type, connected with Stieltjes integration*, *Acta Math.*, **67**, 2, pp. 251–282.

Index of Symbols

Index

Printed in the United States
by Baker & Taylor Publisher Services